# LETTERS TO VANESSA

# LETTERS TO VANESSA

## ON LOVE, SCIENCE, AND AWARENESS IN AN ENCHANTED WORLD

*Jeremy W. Hayward*

SHAMBHALA
*Boston & London*
1997

Shambhala Publications, Inc.
Horticultural Hall
300 Massachusetts Avenue
Boston, MA 02115
http://www.shambhala.com

9  8  7  6  5  4  3  2  1
First Edition
Printed in the United States of America

⊗ This edition is printed on acid-free paper that meets the
American National Standards Institute Z39.48 Standard.

Distributed in the United States by Random House, Inc.,
and in Canada by Random House of Canada Ltd

Library of Congress Cataloging-in-Publication Data

Hayward, Jeremy W.
Letters to Vanessa: on love, science, and awareness in an
enchanted world/Jeremy W. Hayward.
p.   cm.
ISBN 1-57062-077-6 (pbk.: alk. paper)
1. Spiritual life.   2. Imaginary letters.   3. Hayward, Jeremy W.
I. Hayward, Vanessa.   II. Title.
BL624.H377  1997          97-2051
291.4—dc21                CIP

# Contents

# Contents

# *Preface*

The ordinary world is already enchanted. The enchanted world is not a fantasy or a hope for the future; it is *real*, and it is *now*. What keeps us from seeing the enchanted world—*really, now*—is the Dead World story we tell ourselves and each other. We soak up this story unconsciously as we grow up. It comes from a narrow, poverty-stricken vision that our world is made only of lifeless matter. This story was invented over the past few centuries in the name of science. And this vision is still alarmingly effective. It still drives our culture.

In thirty years of practicing and teaching meditation, I have found that unconscious belief in the Dead World story can be a deep obstacle when people try to practice, whatever their specific practice may be. And it keeps many people from even realizing that meditative practice can help their lives on a very practical level.

Ten years ago I wrote two rather technical and academic books to show how science could just as well uphold the enchanted world as deny it. The Dead World story was told for complex religious and political reasons, not purely scientific ones. For some years many friends, especially my wife, Karen, had been encouraging me to try to explain about the two worlds in clear language without jargon. I too felt that it is important for people not just to *believe* in the enchanted world as a new form of blind faith or wishful thinking, but to understand how our deep conditioning obstructs our vision of that world and how we can clarify that vision.

After struggling for two years, trying to write a simple version, I went on a three-week writing and meditation retreat. Before the retreat I had been particularly concerned about the way the Dead World story was affecting my daughter, Vanessa, and her friends. And I had also been deeply touched by conversations with a young friend, Adam, who was in the first year of college. He impressed on me how terribly despondent many of his generation are at the state of the world and the hopelessness of what they see in their future. On the first morning of my retreat I decided to write to Vanessa about these things. At last this book began to take shape.

When people practice meditation without realizing their deep conditioning into the Dead World, their spiritual practice and beliefs can often be just a pretty coat or a new suit or a face-lift that makes them feel better but doesn't fundamentally change anything. One purpose of this book is to show you who meditate, in whatever tradition, how to work through the conditioning of the Dead World story back to a deeper level of feeling and awareness.

Alternatively, many people feel deep spiritual longing for the enchanted world, but when they hear that meditation might help them to see that world, they think that it is all nonsense, because *science says so*. *This* is nonsense, and terribly sad. So a second reason for writing this book is to show you who appreciate the world of science and still long to know the enchanted world a way to live in both worlds as one.

Saddest of all, young people of Vanessa's generation are growing up deeply hurt, depressed, and lost. They see the Dead World, they hear about only the Dead World from people who should know. Yet in their coffeehouses they talk about something completely different. They know the Dead World is not all there is, but they have little idea how to find anything beyond it. And so many take to hard drugs, or even suicide, in the search for something more real. That is why I have addressed these letters to my daughter and her generation: to show a way to go forward and celebrate an enchanted world.

Yet the Dead World is only a tiny fraction of all that is. The world is already enchanted—*really, here, now!* Every tree, every rock, every star, and space itself has awareness and living energy. We have the ability to feel this. And there *are* patterns of energy that can be clearly felt but usually not seen. Call them gods, daimons, fairies, angels, *dralas*, even meaningful coincidence; call them what you will. And *that* story of unceasing awareness could have been told to us as we grew up *along with* the tiny, narrow story of the Dead World. And it too could have been told in the name of science. That is the message of this book.

Please enjoy this book of letters to my daughter. Don't take it too seriously. May it cheer you!

# Acknowledgments

❦

My great appreciation goes to all the friends who read the letters and commented, criticized, suggested; to Emily Hilburn Sell, editor and friend, for guidance and patience; to my wife, Karen, who insisted for many years that I should write the stories of the Dead World and its enchanted counterpart so that people could *understand* them, and who was there with unwavering good heart through the highs, lows, and in-betweens; and to Vanessa and her friends without whose genuine longing, needless to say, these letters could not have been written.

Permission is gratefully acknowledged for use of the photographs of fractals, in Figures 8–13, from *The Beauty of Fractals*, by H-O Peitgen and P.H. Richter, from: Springer-Verlag NY Inc., 175 Fifth Avenue, New York, NY 10010. © H.O. Peilgen, H. Jurgens, D. Saupe.

# LETTERS TO VANESSA

# LETTER 1

$\maltese$

# The Living World We Felt
# as Children

*Dear Vanessa,*

When we are young, we experience the world, *our* world, as magical, alive, sacred. But as we grow up and go through the conditioning of schooling, television, listening to grown-ups' conversations, reading popular magazines, and so on, we gradually learn to ignore the magic of the world, to forget it, and even to deny it. We pick up, almost unconsciously, the notion that the magical world is childish and unreal and that we should put away childish things.

My purpose in writing this little book of letters to you, Vanessa, is to try to show you that there *is* a living, magical dimension to the world we live in. The magical dimension of our world is real, as real as the itch on your arm and the hand scratching the itch. It is the dimension that you and I felt as children but could not describe. We could not describe it because none of the adults around us ever spoke of it, so we did not have any language to describe it. I am going to show you that you don't have to choose *between* the sacred, magical world of the child and the world of the modern adult. You can live in both worlds, for they are one.

I saw your official education in the story of the Dead World begin when you were in the third grade. You came home from school one day and announced with a tone of excitement and seriousness, "Daddy, we learned what matter is today."

Dreading what I was to hear, I asked, "Oh, and what did you learn matter is?"

You replied, "Matter is the stuff the world is made of."

Matter is stuff. Matter has no life or feeling. There's no mind in matter. We all learn that at school. So I became sad and angry, knowing that you too would gradually learn not to feel the mind and heart of your world. How different that Dead World made of matter-stuff is from the world you experienced when you were three years old! Although you didn't explain things logically, you expressed a living, feeling quality in your world. You seemed to feel connected with an alive and loving world.

Once we were driving home one moonlit evening in our old beat-up Checker cab, you strapped in your car seat on the front seat between Mom and me. As the moon crossed the telephone wires you said, "Look, the moon's falling," because it actually did appear to be falling. As we arrived home, you said, "Look, the moon came too." You were so glad that the moon had come with us.

And you loved that huge oak tree on the sidewalk outside our house. It had great branches that spread out over the sidewalk and into the road. You used to play with your dolls under it for hours. You felt that it gave you friendship and protected you. One day, while you were at preschool, the roadworks men came by and pruned it so that it lost almost all its big branches. It looked almost dead. When you came home, you stared at it in astonishment and then cried and screamed inconsolably, "They hurt my friend."

Of course, children do make up stories to explain things in terms they understand. The moon was not falling in the same way a stone falls if you throw it up. And it did not "come too" in the way our friends might have if they had been following us in their car. Yet the feeling you were expressing was true. The moon did, in a way, "come too," because it had never left us. And the *feeling* of the reliability and friendliness of the moon, and of the whole world, was what you were trying to express.

Like you, I too lived in a world full of feeling and meaning when I was a small boy. The world was magical. I don't mean that I believed in Santa Claus and magic wands that could change geese into fairies. My mother, your Granny, was a Protestant Christian (a Congregationalist), and my father, Grandpa, was a civil engineer, an expert in concrete. So they did not think it proper to encourage children to believe in invisible beings of any kind. I felt the world around me was magical because it glowed with feeling, with goodness, with life. And I felt that life and feeling were in the *world*, not just in me.

My bedroom window looked west across a wheat field. I remember my mother putting me to bed one summer evening when I was about six years old. I looked out my window as the setting sun shone on the golden, glowing, ripe wheat field. I remember feeling at that moment a quietly joyful feeling of loving the wheat field, a feeling that filled my body and was so clear that I remember it even now, fifty years later.

Between the ages of six and ten, I felt that everything in my world had a living, glowing quality to it. I used to love to wander around the large garden in the back of our house in England. It was quite a beautiful garden in places, with rose beds and perennial flower beds. In other parts it was shaded and uncultivated. In yet other places, where potatoes and carrots grew, it felt very pragmatic. Those parts felt like Monday morning, which was wash day, when my mother and grandmother used to spend half the day doing the laundry in a crazy old tub that clanked and groaned as it turned.

I can recall the feeling of each part of the garden as, in my imagination, I walk around it now. The living quality of each part of the garden is lodged, not just in my visual memory, but in my whole body. I *feel*, in my body as well as my mind, the cool dampness of the southeast corner, under the fence and shaded by the house. Rhododendron bushes grew there, and there was a slight feeling of foreboding and loneliness. I *feel* the sunny raised area farther down the south side, on which there were rose beds and a paved area for deck chairs. Farther down still was the bright, colorful, and happy-feeling perennial bed. Then came the large area for vegetables, surrounded by low black currant and gooseberry bushes, which a had rich and inviting quality in the summer but felt deserted and forlorn in winter. And so it went, all around the garden.

Well, as I grew up, I also began to learn that the world is made of lifeless, mindless *stuff*, and all the feeling of living connection with the world was buried. I was taught that competition and aggression were the only way to succeed in this life because "it's a jungle out there—red in tooth and claw." I forgot that I had a body. Oh, if someone had asked me, "Do you have a body?" I would have said, "Don't be stupid; of course I do; you can see it." But I came to live so much in my head that I forgot to *feel* my body, to be *in* my body. When we are six, or eight, we feel our body all the time, and we feel the world through our body. We might not realize this until much later, after we have lost it and then regained it. And when we regain it, we regain the memories that went with it, and the feeling of a living world.

When I was seventeen I fell in love with physics after reading *The*

*Mysterious Universe,* a book written in the 1930s by a major physicist of that time, James Jeans. Jeans said that twentieth-century physicists were not just discovering mindless matter but were reading the mind of the universe. Jeans suggested that the "matter" of the universe was more like condensed thought than dead stuff. For a moment, as I read Jeans's description of physics and thought about it, I felt again the wonder and joy of being a child in the world, of feeling the world to be living and to be radiating life and affection to me.

I remember the moment I discovered quantum theory, reading that book. I was sitting by an apple tree in the garden I loved so much. It must have been early spring because the apple tree was in full bloom, and I looked up at the blue sky through the apple blossoms and felt again, for just a moment, some living, feeling, presence all around. And from that moment I decided to study physics at university. I walked around muttering to myself, "Quantum theory, quantum theory," which did not endear me to the girls.

My love for quantum physics was a love for the world and a longing to rediscover the livingness of the world. But once I was at university, I became disappointed, lost heart, and lost my way. It became clear to me that physicists by then believed that physics proves there is *no* mind or feeling in the universe, rather than discovering it there, as Jeans had.

Still searching for the living world, I turned to molecular biology. It was a tremendously exciting time. There was an almost religious fervor that the "secret of life" had been found to be nothing but chemistry. I went on for four years trying to find life in biology, but the living energy of love and feeling seemed to have been taken even out of the study of life.

Now I began to question, "What is consciousness?" and I read a ton of spiritual books—Ramana Maharshi, Krishnamurti, and so on—at this time. At the same time as finding some glimmer of a new vision in these books, I felt increasingly angry and desperate. That vision seemed to be dead and gone in the world I was in. I suppose, looking back, that I was on the edge of a deep self-hating and world-hating depression. I believe a lot of your friends may be in this state of mind right now, Vanessa. Please tell them, "Don't give up!"

Then something magical happened: I found a genuine living spiritual tradition, in the form of the teaching of G. I. Gurdjieff. Gurdjieff was born in Russia around 1877. As a young man, he disappeared and traveled in India, Tibet, and the Middle East. He went "in search of the miraculous," that is, in search of something that could nourish *all* the possibilities for human life. He appeared again in Moscow in 1912, and for the next thirty-six years he taught, first in Russia, then in France and America. He called his teaching the Fourth Way, or the Way of the Householder, to distinguish it from religious approaches to spiritual training.

Meeting the Gurdjieff world in 1966 was a powerful event in my life. After I had been with the group for about a year, I attended a group meeting in the living room of one of the members, an ordinary, middle-class living room. The group leader started speaking, and I looked across at him. Suddenly the room changed. The space became full and rich and the face of the person speaking was dancing and glowing in that living space. I was *in* that space, and the space flowed through me and brought me a feeling of quiet joy and intense vividness. For the first time since I was a small boy, I clearly *felt* that there is a dimension to our ordinary existence that is profoundly rich and alive. And as we grow up, we learn step by deliberate step to deny this reality and ignore any hint of it in our life. This discovery was not just something new to think and believe but something that I also *felt* deep in my body and heart.

In 1970 I met Chögyam Trungpa Rinpoche, a young Tibetan Buddhist teacher. Trungpa Rinpoche had escaped from Tibet in the late fifties and had gone to England to study at Oxford University. There he learned to speak fluent English and began to teach Buddhism to English people. After a few years he realized that his monastic robes were an exotic fascination for people and distracted them from hearing the simple teaching of Buddhism. So he exchanged his robes for ordinary Western clothes, and he also married an Englishwoman. Rinpoche had just arrived in America with his new wife when I met him. He was teaching a weekend workshop entitled "Work, Sex, and Money"—from the meditative point of view.

When I first met Rinpoche, I immediately felt that he lived constantly in that profound dimension that I had discovered in the Gurdjieff work. At the same time, he lived an ordinary human life like mine and my friends' (many of whom were "hippies") and had a tremendous sense of the humor and silliness of life. I knew then that the magical dimension of life is real and livable. And the Tibetan Buddhist teachings showed me a way I could reconnect with the sacredness for which I had been searching: by looking directly and honestly at my own mind and experience.

Our modern society too has largely lost touch with sacredness. Yet many people do experience their world as a living being. For example, the poet Kathleen Raine writes:

> There have been times—or shall I say there was one unforgettable time—when I saw this world as a living being. It was a very simple event, such as might happen any day, to anyone. I was sitting alone in my room, writing at a table on which there stood a hyacinth growing in a glass. I was thinking of nothing in particular . . . when before my eyes the world changed. The hyacinth appeared in a flow of living light that was in some mysterious way not separate from me but like a part of my own being. Inner and outer were indistinguishably one. At the same time I know that I was seeing more fully what was really there. I have remembered it always, although I have never again experienced the like. But once is enough to know forever.

In our modern world we rely on a narrow, distorted idea of science to tell us the story of the stuff the world is made of. We look to this pseudo-science for the "truth," to tell us what is real and what is not real. Much of what scientists say is tremendously helpful and true. It is not that the world that their stories describe is completely wrong, or doesn't exist at all. It is just that the modern description leaves out so much—it leaves out the sacredness, the livingness, the soul of the world. And it does get troublesome when some scientists tell us, often with a voice of authority, that the part they leave out is really not there.

The way we feel and experience our world is deeply affected by what we believe about it—by the stories we tell ourselves about it. Scientists tell us the stories that make our world, the "modern world." Generations of misunderstanding and misuse of science have brainwashed us, so that by the time we are seventeen most of us don't feel that we are immersed in a world quivering with heart and life. We don't even know that such a world exists. We may not want to admit it, but most of us often feel deadened and hopeless in a universe that we think couldn't care less about itself or any part of itself.

Right from the start I must emphasize that these letters are not by any means anti-science. Science has been used as the voice of authority against the magical world—yet science can equally as well speak *for* that world. The problem is not with science but with the way the story of the Dead World has been told, in the language of science, for the benefit of all kinds of other beliefs: religious, commercial, political, and so on. Many people say, "Oh well, I didn't learn science in school," or "When I studied physics in high school I thought it was bullshit, so this doesn't apply to me." But our conditioning goes much deeper than when we first consciously remember learning science. The conditioning is in everything we see and touch from the moment we enter the world, everything we hear from the moment we begin to speak, everything we read. The Dead World is in all of us. All of us learn to see our world through the story of science from a very young age, whether we know it or not. If we don't know this and feel it in our bones, then we just take for granted the truth of the story of science and live by it. And that is even more dangerous.

People in other societies tell very different stories about the world than the ones we tell. And they seem to experience the world differently, just as their stories tell them to.

The differing stories that various groups of people tell make societies very different from each other. What makes the difference between Japanese people, born and brought up in Japan, and Japanese-American people? Of course there are many other factors involved, but the stories that they were told or soaked up unconsciously as they grew up play a huge part. A Navajo woman brought up among Navajo people is very different from a woman of Navajo race brought up in a city, because they were not told the same stories as they grew up.

The stories we were told as children showed us what to experience as *real* in the world. We usually think that what we believe is real is based on what we see and hear. But this is only partly true. It is just as true to say the opposite: that what we believe is real decides what we will see and hear in our world.

To show you what I mean, let me give you some examples. Charles Darwin, who invented the theory of evolution, sailed around the world in a large sailing ship called the *Beagle*. During his voyages he made many detailed observations, such as recording the different kinds of finches on the Galapagos Islands, for example. So he was a very clear and accurate observer. He was astonished to find that when the *Beagle* was anchored on the horizon near the islands, the native islanders simply could not "see" the ship, even when it was pointed out to them. They could clearly see the small rowing boats ferrying crew between the *Beagle* and the island, which were probably about the same size as their own canoes. But they could not see anything on the horizon that they recognized as a ship. They did not believe a ship could be that size, so they could not see it.

Here is another example of people not being able to see what they do not believe in. Years ago, Harvard psychologist Jerome Bruner conducted experiments on the psychology of perception. In one, Bruner asked students to look through a tube at the end of which he flashed two series of playing cards very fast. (This machine is called a tachistoscope.) One series showed normal playing cards and the other had the colors reversed, so Hearts and Diamonds were black instead of red, and Spades and Clubs were red. The students had a much harder time recognizing the color-reversed cards and often went to great lengths to reinterpret and "regularize" what they saw. One student reported that the red six of Clubs was a *real* black six of Clubs but the light inside the tachistoscope was *pink*. There are many experiments that demonstrate the same basic fact: people normally do not see things that they do not believe in, or do not expect to see.

And what we do believe affects, also, *how* we see (or hear or smell or taste or touch) our world. For example, look at the two pictures in figure 1, on page 9. The lower picture is a line tracing of a painting by van Gogh of his bedroom in the French town of Arles. The top picture is a line tracing of the same room with the "correct" perspective lines. If I asked you which picture you thought most accurately represented the room, your first response would probably be, "The top one." Yet the bottom picture represents how van Gogh actually *saw* the room.

If you look at the room you are sitting in now, you will probably see it as if with perspective lines, much like the upper of the two pictures. But if you try to see it like the lower picture, by relaxing your gaze and imagining that you are seeing it in that way, you might be able to. People once thought that van Gogh was mad, but now we realize that he probably saw the world without the filters that we usually see it through, such as the filter of perspective lines. Perhaps that is why his paintings are so alive and vivid. The colors are so bright, the lines so rich, curving and swirling as if the world he saw was full of life and energy. And this is how he described the world he saw. Van Gogh felt the world deeply. He said, "The nature we see and the nature we feel, the one out there and the one in here, both must permeate each other to last, to live." But when we see the world through our perspective lines, it feels flat and dead.

That reminds me of a story about Picasso, who was asked by a man, "Why don't you paint people as they *really* are?"

Picasso replied, "And how are they *really?*"

The man took out a photo of his wife and said, "Like this."

Picasso responded, "She's awfully flat and small, isn't she?"

People in the Western world did not always see the world as if through a straight-line grid, as we do. This way of seeing was invented in the fifteenth century by an Italian architect, Filippo Brunelleschi. Paintings

FIGURE 1

from before this time have no perspective, and people are crowded together as if they were flattened against a wall. You have probably seen medieval religious paintings like this. Brunelleschi found that he could paint a more lifelike scene if he put a grid of perspective lines between himself and the scene. He then drew the lines on his paper and placed the people and objects between the lines.

This straight-line way of viewing the world became fashionable. Everyone began to paint that way; buildings were built in straight lines; and whereas until then villages had been built with curved streets to protect from the wind, now the streets were built straight. And now we are so used to seeing this way that we feel it is "natural." Yet in nature there are almost no straight lines. Have you noticed how our dog, Sernyi, and our cat, Peter, walk down the lawn? They have made a little path between the deck steps and the lake. And the path is quite crooked!

Here is one more example of how what we believe affects what we can perceive or even imagine, this time in the form of a riddle. A father and his son are driving to a football game. They begin to cross a railroad track at a crossing. When they are halfway across, the car stalls. Hearing the train coming in the distance, the father desperately tries to get the engine started again. He is unsuccessful, the train hits the car, and the father is killed instantly. The son survives and is rushed to the local hospital for brain surgery. The surgeon, on entering the operating theater turns white and says, "I can't operate on this boy, he is my son." The riddle is, what is the relation between the boy and the surgeon? Take a few minutes to think about this one. I will write the answer backward: rehtom s'yob eht saw noegrus eht. Most people, men or women, just don't get it! See how your preconception about who could be a surgeon influenced your imagination of the scene? These kinds of preconceptions create the world we experience *all the time*. And this will be a big part of some later letters.

So this is a powerful principle—what we believe about the world, the stories we carry around buried deep in our body-mind—affect how we feel and experience and interpret what's there. In the next letter I will give you some examples of other stories people tell and other worlds they live in.

# LETTER 2

## Stories with Feeling, Stories with Soul

*Dear Vanessa,*

As I promised yesterday, in today's letter I want to write a little about a few of the many societies that did not lose their experience of the magical world. And I will begin with a world that is really close to home—the world of medieval Europe.

The medieval universe was full, populated by all manner of living, feeling beings, from plants and animals to humans to spirits and angels, all the way up to the mind of the Christians' one god, or down to the mind of the devil. Any being that could be imagined must exist, people believed, out of the generosity of the creator. And this Great Chain of Being stretched from the devil and his servants through plants and animals (which included women!) to man, and beyond man to the angels and to God.

The one god was above but his spirit also permeated all of life. The Christian church allowed and incorporated the pagan feelings of a god permeating and impressing himself on the phenomenal world. The angels were very far from being the chubby babies with wings that we see pictured today, nor were they merely guardians who appear to help people in time of need. Angels were powerful and terrifying, being closer than humans to the one "almighty" God. Some were originally pagan gods and goddesses, co-opted by the church.

What was in the heavenly realm was mirrored in the earthly one.

Hence the well-known phrase "as above, so below." In fact, *divination* meant finding the divine behind all appearances. There wasn't a distinction as we know it between the psychic and physical, or inner and outer, or literal and symbolic. There was a feeling of belonging—a feeling of holistic connection of all phenomenon. People felt connected to something larger than themselves and sought to discover their role in that whole.

At this time, alongside the Catholic beliefs and dogmas, there flourished sophisticated systems of astrology, alchemy, and magic. There were contemplative practices involved in all these traditions which included the study of sympathetic relationship, or resonances. People practicing these spiritual traditions learned to feel the resonances between the various levels of the cosmos: for example, between the solar system, parts of the human body, and various metals or plants. They thought that by contemplation of various connections between things in the natural world they could come to a direct intuition of connections on other levels.

To modern scholars alchemists appear to have been studying how different metals and other elements, such as sulfur, combined. We're told it was nothing but the beginning of modern chemistry. But to the alchemists it was also the study of how different elements of the personality, corresponding to the different metals and other chemical elements, combined. From their working with elements, alchemists felt that they were also transmuting their own spiritual nature.

It seems that in medieval times people were able to access a quality of consciousness that is unknown to scientists and to our general culture today. Some writers call it "participating consciousness," meaning real knowledge of an object that occurs through the union of the subject (the "I") and the object.

Nowadays we believe we can know about an object only by examining it as something separate from us. The point of participating consciousness is that the person *participates in* the object. He or she recognizes the resemblances and correspondences between all things; feeling how all things have relationships of sympathy and antipathy to each other. This was the belief and practice of alchemists who recognized no distinction between mental and material events. As the historian Morris Berman writes, "It is not merely the case that men conceived of matter as possessing mind in those days, but rather that in those days, matter *did* possess mind, *actually* did so." And he asks, "Which is the *altered* state of consciousness? Why is today's view more believable?"

Men and women perceived phenomena in their world that we simply no longer know in the modern world. The medieval world was *actually* magical, it was enchanted. People saw all manner of beings in addition to ordinary humans—angels, ghosts, fairies, and all kinds of spirits of the fields and woods. They not only believed in these beings, but they actually saw them, or at least they believed they saw them. There are many stories of people meeting angels, ghosts, demons, or fairies that are quite soberly written and that we have no reason to believe were deliberate falsifications.

For example, in her book *The Medieval Vision*, historian Carolly Erickson tells us of a monk of Byland Abbey in Yorkshire who set down a number of encounters between the people of his neighborhood and beings of other domains. "One of them" she says, "concerned a tailor named Snowball whose meeting with a disembodied spirit involved him, his neighbors, and the local clergy in a drawn-out transaction with the incorporeal."

"Riding home one evening," Erickson continues, "Snowball saw a raven fly around his head and then fall to earth as if it were dying. When sparks shot from its sides, he knew himself to be in the presence of a spirit and crossed himself, in God's name forbidding the creature to harm him." The spirit attacked Snowball twice more at which point Snowball decided to find out what it wanted. It appeared that the spirit was stuck in some unpleasant realm because of some wrongdoing he committed when he was alive in human form, and he needed a priest to come and absolve him. He made a deal with Snowball, who fell ill for a few days and then went to fetch the priest. And so the story continues. The spirit is finally released and in return he tells Snowball of his future. Erickson concludes,

> Medieval perception was characterized by an all-inclusive awareness of simultaneous realities. The bounds of reality were bent to embrace— and often to localize—the unseen. This perception, where it is alien to modern consciousness, may be likened to an enchantment. Medieval people lived in a perceptual climate in which noncorporeal beings were a familiar and to some extent manageable force, recognized alike in theology and popular culture. . . . Extraordinary appearances—unusual natural configurations, visual portents, dream messages from the dead, divine and infernal warnings, intellectual illuminations, visions of the future—everywhere complemented ordinary sight. . . . Understanding the medieval past means coming to terms with a quality of awareness that much of modern education is intended to discredit. The visionary imagination, long a disquieting embarrassment to rationalistic histori-

ans, was in the medieval period not aberrant but mundane, not un-earthly but natural, even commonplace.

But such experiences of nonphysical beings are not limited to the distant past. W. Y. Evans-Wentz was an anthropologist and religious scholar, and was one of the first translators of Tibetan Buddhist texts. In the early 1900s he spent two years in Ireland, Scotland, and Wales, interviewing elderly folks who could still see and hear fairies. The fairies these people saw weren't the wee little cute ones we see in the storybooks. They were tall, full-sized fairies who looked like human beings but were translucent and luminous and wore antique clothing. These fairies were witnessed in all kinds of activities—religious processions and rites, ordinary hunting activities, helping humans, and so on. These sightings weren't made just by isolated individuals; lots of people saw them.

When Evans-Wentz asked an elderly man on the Isle of Man why the younger people weren't seeing them, the man replied, "Before education came into the island more people could see the fairies; now very few people can see them." This education most likely told them that fairies don't exist, so, since they weren't allowed to see them, the young ones lost that potential and then no longer did see them.

Yet there have been surprisingly many sightings of fairies in modern times. Sometimes they may be scary, and on other occasions benevolent. For example, a friend of a well-known British folklorist, Katharine Briggs, told her that she had been suffering from an injured foot and had sat one day on a seat in London's Regent's Park. She was wondering how she would find the strength to limp home, when suddenly she saw a tiny man in green who looked at her very kindly and said, "Go home. We promise that your foot shan't pain you tonight." Then he disappeared. But the intense pain in her foot had also gone. She walked home easily and slept painlessly all night.

I taught a program in France some years ago at an old chateau that had been converted into a Zen center. It was a very intense time because not only did we talk, we also practiced and invited the "gods" to join us. There was a young German woman there who, halfway through the program, started to see a man dressed in green medieval clothes looking through the window. She knew that he was not human, and she became frightened. She said that she had seen these beings when she was a child but had tried to stop seeing them, and had not for twenty years. I encouraged her not to be afraid and to ask the gentleman what he wanted. Toward the end of the program she told me that he had told her that he was a protector of the land and the chateau. He told her he was just watching to be sure that we and the Zen people treated the land with care.

☙

Like the world of medieval Europe, the Navajo world, too, is alive. It is permeated through and through with subtle life and power. The Navajo people speak of Holy People, *diyin dine'e,* who are the living heart of all things, human, animal, plant, and inanimate. Every visible object in the world has an invisible aspect that is its *diyin dine'e.* There are Mountain People, Star People, River People, Rain People, Corn People, and so on. Natural phenomena such as mountains, mesas, canyons, caves, rock formations, great rivers, weather conditions, and light are the dwelling places of the Holy People.

"I was brought up to believe that they were the soul of things," Navajo artist Baje Whitethorne says of the Holy People, "the soul of all things, all living things. Just like God would be the soul of things, he is around everywhere, so, in Navajo tradition, in our religion, they say the same thing. They are the soul of things. . . . They were in the rocks or in the trees, or just about everywhere."

"When I go out at dawn, I pray to the breeze, to the newness of life," says Kalley Musial, a Navajo potter, "not to anybody in particular but to the birds, the plants, just to life itself. I pray to the dawn, which is the awakening of life—for the plants, for the birds, for us.

"When I pray at noon, I pray to the sun, which gives us warmth and life and growth. When I pray in the evening, I pray to the breeze again, to what is around me, to air, to what comes with the evening. We pray to all, to everything. It's just like God is out there, the essence of life, air, rain, everything."

The Holy People are, in many respects, much like the ordinary Navajos. They are human in appearance and live much as Navajos do. But they don't have bodies of matter. Their bodies are more like wind and light or, as the Navajos say, Holy Wind.

*Wind* symbolizes the universal living energy that pervades and enlivens all things. These living energies cannot normally be seen but can be observed only by their effects, just as the physical wind can be seen only by the motion of branches on a tree or clouds in the sky. In humans this subtle, not normally seen, aspect is called the Wind within one. The Holy Winds of each separate thing are not really distinct. They all are really part of One Wind and the living energy of Wind flows in and out of even the most apparently solid objects.

— 15 —

According to Peter Gold, author of *Navajo and Tibetan Sacred Wisdom*, "Holy Wind is a glittering, pulsating, breathing fusion of all the animating and enlivening energies of a living cosmos. It is the power behind the Universal Mind [awareness] permeating all the elements and phenomena of the cosmos." Together, Holy Wind and Universal Mind compose the state of being known as *ho'zho*. Usually translated as "beauty," ho'zho also means harmony, happiness, health, and balance.

The difference between Holy People and us is that the Holy People live completely within ho'zho—they are completely at one with the forces, rhythms, and inherent order of the cosmos. We too can develop our imperfect body-minds to the state of ho'zho because we are of the same stuff as the Holy People. We are all emanations of the oneness and power of ho'zho that permeates the earth and everything on it.

To the traditional Navajo people, living according to ho'zho, is living with balance, peace, and beauty as the goals in life. That is the aim in day-to-day actions and prayers. As Navajo artist Jimmy Toddy says, "Every prayer you start with, 'Beauty before me, beauty around me, beauty ahead of me.' Ho'zho—the prayer goes that way. Every prayer you say with that, beauty, beauty."

In Japan, Shinto is the way of the *kami*, which is usually translated as "gods." But Shinto is a way of human life, according to Sokyo Ono, a Japanese authority on Shinto, "an amalgam of attitudes, ideas, and ways of doing things that through two millennia and more have become an integral part of the way of the Japanese people." He describes the term *kami* as an honorific term for noble, sacred spirits. It implies a sense of respect, love, and awe. He says, "All beings have such spirits, so in a sense all beings can be called kami or regarded as potential kami."

In the Shinto world, like the Navajo, there is no, one all-powerful god that is the creator and ruler of all. The world is self-created, and this self-creation comes about through the harmonious cooperation of the kami as each performs his or her particular mission. The kami are found in the qualities of growth, fertility, and production; in natural phenomena such as wind and thunder; in natural objects such as the sun, mountains, rivers, trees and rocks; and in some animals. Kami are the guardians of the land. They are the heart energies of occupations and skills. They are the spirits of ancestors, national heroes, men of outstanding deeds or virtues, and those who have contributed to civilization, culture, and human welfare.

How do the Japanese experience their kami? Sokyo Ono writes, "The Japanese people themselves do not have a clear idea regarding kami. They are aware of kami intuitively at the depth of their consciousness and communicate with the kami directly without having formed the kami-idea conceptually or theologically. Therefore it is impossible to make explicit and clear what is fundamentally, by its very nature, vague."

This is illustrated by a story told by mythologist Joseph Campbell. He was in Japan for a conference on religion and overheard another American delegate, a social philosopher from New York, say to a Shinto priest, "We've been to a good many ceremonies and seen quite a few kami shrines, but I don't get your ideology, I don't get your theology." The Japanese paused as though in deep thought and then slowly shook his head, saying, "I think we don't have ideology, we don't have theology—we dance."

In Japan, the countryside is filled with small shrines to the kami, placed at power spots of the land. Every garden and home has at least one shrine marking the power spot. The shrine is not elaborate; it could be nothing but a rope or a group of rocks marking off an area. Or it could be a small wooden dwelling with an opening in which to place fresh flowers.

Most older Japanese, whether their actual religious worship is Buddhist or Christian or Shinto or none at all, have respect for the kami. They feel their presence and understand the need to communicate with them in order to maintain the proper flow of energy in their world. Even a businessman building a bank at a particular location will perform the appropriate ceremonies to pay respects to the kami of that place before beginning building. And ceremonies will be performed, throughout the building process, to gather the energy and power of the kami.

The pre-Buddhist native traditions of Tibet speak of energy-presences, called *drala*s, very similar to the Japanese kami, the Navajo Holy People, or the pagan gods and fairies of medieval Europe. Drala literally means "beyond enemies." And the enemy is aggression and territoriality—anything that divides our world into separate parts that fight each other. So drala energies harmonize the parts of our world and heal the fragmentation of it. Drala is a new term to us, and does not come along with all the rigid and degraded associations of terms like "gods," "fairies," or "angels," and so on, so I will use it quite frequently in these letters.

Chögyam Trungpa, founder of the Shambhala sacred path of warriorship, felt that, although there has been great development of wealth in the Western world, much of the vitality of the land has been harmed through manufacturing, mining of the earth, and so forth. Because of this, the dralas have departed. To restore vitality and heal a wounded situation, he taught Shambhala practices to people in the modern world, for them to reconnect their own heart wisdom with the energy and power of the dralas. These practices, he said, could return brilliance and dignity to our physical world and body, potency to our speech, and courage and strength of heart to our minds. He emphasized that we can actually connect our own being with the dralas; they are not merely comforting fictions. But this connection will happen only if we actually practice, not just talk about it.

⍟

In all these ways of living—whether it be medieval Europe, the Navajo peoples, the Japanese Shinto, or the Tibetan drala teachings—people seem to experience multiple dimensions to their world. There is the ordinary realm of material reality, and there is the realm of gods, spirits, ancestors, and angels. And these two are simply different ways of perceiving and experiencing the same one world.

For example, Carolly Erickson writes of a thirteenth-century manuscript which tells the story of three monks who agreed to journey together to "find the place where heaven and earth join." Their long journey is described in factual geographical detail as they cross the Tigris River, travel through Persia (now Iran), and into the plains of Asia. On the way, they meet all kinds of strange beings: people less than two feet high, a bleak mountainous region filled with dragons, still higher mountains populated with elephants, a place where sinners were undergoing terrible torments, and so on. Erickson writes,

> The account of the monks' journey, which is found in a manuscript devoted to serious geography, combines several dimensions of reality into a single and continuous landscape. In this account, the spiritual geography is localized . . . and is treated as a part of terrestrial geography.
>
> The multiform reality which forms a backdrop to the monks' journey may be likened to an enchanted world in which the boundaries of imagination and factuality are constantly shifting. At one time the observed physical limits of time and space may be acknowledged; at another time they may be ignored, or from another point of view transcended. Yet so constant and so automatic is this expansion and contraction of the field of perceived reality that it goes unnoted and unreconciled by medieval writers.

The Australian Aborigines, too, believe that their universe has two aspects: the ordinary physical world in which they live and another world called the Dreamtime. They actually *see* both the ordinary world and the Dreamtime as equally real. The gods of the Aborigines, who are also their Ancestors, continually walk through the land *now*, chanting stories. The land, which otherwise would be flat and dead, is continually brought to life, moment by moment, by the telling and retelling of the stories. The stories make the hills and valleys and rocks and pools. The songline or dreaming-track of an Ancestor is the path that he or she takes in forming the land.

Traditional Aborigines devote their lives to learning about and obeying the rules of the Dreaming. They learn level after level of interpretation of the songs and stories. And the more they learn, the more they see in the land itself. The land itself is the textbook through which Aborigines are educated. Information about every dimension of existence is hidden in the stories of the land. To learn the secrets of the land is to learn everything worth knowing.

The songs and the songlines are so important because, at the same time as describing reality, they are the actual forces that make it work. The songs are the cosmic rhythms and melodies that give the everyday world its form. They were not composed by humans, for then they would not have the power to hold the external world together and bind it to Dreamtime. The songs come to the Aborigines from the Ancestors. They are handed down through generations and continually renewed through dreams. When an Aborigine walks a songline and sings that song, he becomes part of that Ancestor and takes part in the continuing creation of the land. Perhaps the Navajo would say, "he walks in ho'zho."

The land lives in this way, and holds the wisdom of Ancestors, for many traditional peoples. Cree writer and university professor Stanley Wilson writes of his experience at an education conference in Georgia. He was standing on the campus, speaking Cree with his wife, Peggy, when he suddenly felt a feeling of elation that he had never before experienced, followed by a debilitating depression. When later he asked an elder about this experience, the elder told him that the land held ancient memories of his ancestors also contained within the biological cells of his own bones. These ancient memories were triggered in him as he walked on that land.

To understand this story, you should know that the campus on which Wilson walked was on the Trail of Tears, so called because it was the trail thousands of Native Americans were forced to walk—many of them dying along the way—when they were evicted from their homelands in Georgia and relocated on reservations in Oklahoma. Wilson's ancestors first expressed delight and elation at hearing him speak their own tongue, but then communicated to him the sorrow of that land.

🙰

What could all these stories mean to us, Vanessa? Do the gods, angels, fairies, spirits, Holy People, Ancestors, kami, dralas, actually exist? Well, that may not be as simple a question as it sounds. They do not exist as totally separate beings. The forms they take clearly are determined by the cultures that experience them. But they are not merely subjective imaginary beings either; that is, they are not just "all in your head." What I mean by this will become clear as we go on, I hope, as we examine the nature of experience and how our so-called "real" world comes about.

I will show you in later letters that we can say the same of everything that you believe exists in your world, trees, rocks, birds, our dog Sernyi, Mom, and me: none of these are fundamentally separate from you, yet nor are they merely your subjective imagination. It is just the same with drala and so on. And because we are not fundamentally separate from them, we can connect with them and draw their energy into our life. And because they are not "just in your head," they, in turn, can empower us and aid us.

Among all the peoples that I have written about in this letter, as well as most other native peoples around the world, we find there is a universal theme: the world is alive, permeated with living energy, responsive feeling, and awareness. And everything we see and hear and touch in our world partakes in that responsive, living energy-awareness.

We might say that native peoples are connecting with the soul of the world—not soul in the sense of a separate "thing" we are each supposed to have, but something much more real. As Thomas Moore says, in *Care of the Soul*, soul is "not a *thing*, but a quality or a dimension of experiencing life and ourselves." It is a quality that permeates everything, like the Holy Wind of the Navajo. It is a pulsation, a vibration, as the heart vibrates when we see something beautiful or ugly. Soul is some unnameable depth of feeling, a cherishing of things, and relating to things with heart and mind together; and in it there is a longing to join ourselves with things—a longing for the participating consciousness of medieval times. Through it we know the essences of things, because the soul in us vibrates sympathetically with the soul in things.

The living, responsive feeling-energy-awareness that pervades everything is the common thread of all the various ways people speak of invisible but knowable beings, though the actual details of the stories may vary tremendously. Their stories also tell how they communicate or dance with those living patterns of energy-feeling-awareness. I have especially emphasized in this letter the medieval tradition, because it is very close to us, it is only just beneath the surface of the modern veneer. We should remember that all of these beings, even the fairies or angels of medieval times, are not at all cute and babylike as most stories and pictures of today would have us believe. They had (but I should say *have*) tremendous power. So we don't necessarily have to go to foreign lands or peoples to rediscover the power and sacredness of the natural world. We do have to open our eyes, minds, and hearts, right here.

# LETTER 3

# *The Story of the Dead World*

*Dear Vanessa,*

As I was preparing to practice mindfulness meditation this morning, I glanced out the window across the snow-covered meadow, through a gap in the dark pine trees. I looked across the frozen white lake to another row of pine trees way off in the distance on the far shore of the lake. It could have been a rather gloomy scene—all white and black. But as I looked, a deep red glow appeared just above the trees on the horizon. The glow increased until it was the tip of a golden circle, and as I kept watching, the circle gradually rose above the line of trees. I felt a quiet, deep joy and said the words, "The sun is coming up." As I looked at the sun and the trees and the lake, there was a warmth, a livingness to it all. My feeling went out to it, quiet, soft, gentle feeling. And the sun and the lake and the trees met my feeling. Something there vibrated, resonated with me. We were there together.

It was the same quality of feeling that I had last spring when I planted roses in our perennial bed. Do you remember how I came home with three rosebushes, and I was so delighted, cradling them in my arms as if they were my newborn babies—thorns and all?

When I planted them, I had never planted roses before, and I didn't really know quite what to do. So I asked the roses. And whenever I went over to that part of the flower bed I felt something coming to me from the roses. What was it? I can't say, but I do know that *something*, some *feeling*, radiated to me from the roses and from me to the roses. We communicate at the level of feeling. And it is the same level of communication, feeling communication, or feeling-awareness, or soul communication, that I can have with our cat, Peter, or with you or Mom, when we are sitting quietly not speaking. Of course, the communication that I have with you or Mom is more complicated than that with the roses. And if we start talking, we very quickly lose that level of communication, because it's very difficult to put into words. Words speak mainly to the surface of things and can't convey deeper levels of feeling, except perhaps in the best poetry.

Now, to return to the lake and the pines outside my window, or to the roses in our garden. When I pay attention to that level of feeling-awareness, there is an openness in me. I feel that openness at the level of the chest, the heart. It is as if I am looking at the trees or the roses through my heart instead of through my brain, as I so often do. And when I look in that way, I feel as if something from the trees, or the roses, comes back and enters me. And I give something from myself, from my heart, to the roses.

There is an actual transfer of a very fine kind of something—a feeling-energy, we could call it. And this transfer of feeling-energy can only happen if I and the roses are in harmony with each other. We are resonating with each other. This is very analogous to what happens if you play a note on a guitar and there is another guitar leaning against a chair nearby. The second guitar will start to sound the same note as the one you struck on the first guitar. This is resonance. And there is actually a transfer of energy from the first guitar to the second, and then back from the second to the first. This principle of resonance runs throughout the universe, and I'll be writing quite a lot about it in these letters.

❧

In our culture, a lot of people have feelings such as these, or other even stranger feelings they can't understand. They may experience what a friend is feeling or thinking who's many miles away. Or they may feel the presence of someone they can't see standing near them, or they may actually see this presence. People report all kinds of peculiar things. Often they don't know what to make of it, and keep quiet. They don't want people to think they're crazy, after all. We may deny or ignore experiences that are important to us because they don't fit with what we've been told. More and more, though, some are willing to talk about these things. You see best-sellers about near-death experiences and shows on Discovery Channel and PBS describing encounters with angels. There's always the requisite scientific debunker appearing to talk about "hallucinations" and such—to say, "It's all in the head."

We certainly must be wary of people's tendency to exaggerate and elaborate. But unseen energies do permeate our world. And when our perceptions become more open we often become more aware of these energies. When you practice meditation, sooner or later you are going to feel these energies, even though that is not necessarily the *point* of meditation. For example, you might feel the insubstantiality of things or the energetic quality of space or the life in a rock. You may feel yourself profoundly connected with the world in a physical way. You may also have a hard time believing your experience is real, and trusting your experience. A deep-rooted doubt says, "I know this *cannot* be real, because the scientists say so."

We customarily believe that scientists are finding out the objective truth, free from prejudice and wishful thinking. A scientist is interviewed on TV: "We now know," she says, "that there is a virus responsible for cancer of the colon. We do not yet have a cure for this cancer, but we expect to within ten years." Another scientist reports, "We have just discovered a new elementary particle called a top quark. We now know all the particles that exist in the universe and are close to having a complete theory of everything."

Francis Crick, a Nobel Prize winner, says, "You, your joys and sorrows, your memories and ambitions, your sense of personal identity and free will are, in fact, no more than the behavior of a vast assembly of nerve cells." Most of us would take these reports deadly seriously. We don't question them, we just let them infiltrate our lives. Even people who see themselves as basically intuitive are profoundly affected by the scientific worldview that permeates everything.

If anyone wants to support an opinion all he or she has to say is, "Scientific studies have shown . . . ," and that more or less puts an end to any debate. We are encouraged to believe in quarks, a particle more elementary than the electron (have you ever felt or seen a quark, by the way?), and to believe that they are as real as robins or rocks. But we may not believe in precognition or ghosts without being considered a little stupid, even if we have actually seen one, because scientists tell us they do not exist.

Do you remember that scene in the Monty Python movie, *The Meaning of Life*, where the chaplain in a boys' school is leading the boys in morning prayers. He says, "Dear Lord. Ooooooh, you are so BIG. So absolutely huge. Gosh, we're all *really* impressed down here, I can tell you." Well perhaps instead of God, we should now be saying this to a statue of The Scientist. I was trained as a scientist, as you know, Vanessa, and I have tremendous respect for true science. But when science becomes the dogmatic religious authority, then, I think, some scientists feel sad and slightly nauseous.

Let's look at the story we've been asked to believe, the story that was told to us over and over again, in third grade, fourth grade and all the way up; the story that we read in magazines and see on TV every day of our lives.

One of the major views is that the human being is a machine—not *like* a machine, *is* a machine. When your mother took that university course in human physiology this year, the very first paragraph in her very thick book said, "The *mechanist* view of life holds that all phenomena, no matter how complex, are ultimately describable in terms of physical and chemical laws. In this view, which is that taken by physiologists, the human being is a machine—an enormously complex machine, but a machine nevertheless. . . . The mechanist view has predominated in the twentieth century because virtually all information gathered from observation and experiment has agreed with it." So you find what you're looking for based on how you set up the experiment to look for it. But we'll talk about this later.

Let's make a list of some of these major views. We can call it

### THE STORY OF THE DEAD WORLD

- Matter is the stuff the world is made of; it has no life or mind (awareness, consciousness, or soul) *at all.*
- You are nothing but a complicated lump of matter.
- Outside of your body there are other lumps of matter we think of as other people.
- Mind, awareness, or consciousness are nothing but products of brain electricity and chemistry.
- Your awareness, consciousness, feeling, and sense of self begin when you are born. When you die, all these come to an abrupt end—absolutely and completely. Nothing continues.
- Between your birth and death you live in a fundamentally lifeless, unaware, and unfeeling world, a world without soul.
- Time exists as an ultimate reality completely outside of you and

goes on without you—a line stretching from infinite past to infinite future.

- Nature is governed by the struggle for survival of every living being against each other.

- Because of this fundamental struggle in nature, self-interest is the motivation for all human and animal behavior. Altruism—caring for another before yourself—is an illusion.

- Your feelings that you have free will, that you can choose to behave selfishly or kindly, are all delusions generated within the brain.

- Any "meaning" that you feel is nothing but your subjective projection.

- Any experiences that you have, or anyone else says they have, like precognition, psychokinesis, telepathy, or out-of-body experiences can only be hallucinations or hoaxes.

- Entities such as "ancestors", "gods," "angels," kami," "drala," and so on, do not exist. They are nothing but inadequate attempts to explain and control the natural world before science found the true explanation.

- The idea of an "inner" or "spiritual" life is delusion, though it may provide psychological comfort to the weak.

This is what is assumed behind almost all popular, mainstream news and magazine stories about the latest discoveries in space, medicine, or human behavior. It is the modern catechism—what you are expected to believe as an intelligent, educated, "grown-up," adult of the modern world. If you protest that it doesn't feel right, that you feel intuitively that there must be more to the universe than this, you will be told to stop being weak and irrational and face the facts.

❦

We live as if our bodies were isolated lumps of stuff; therefore, we lose our health-giving connection with the earth.

We live as if we existed in dead, empty space; therefore, all our energy and insight must come from within, and we constantly feel overcome with anxiety lest our energy run out.

We live as if time did indeed flow from past to future; therefore, we do not rest in this moment at all.

We live as if our minds were located somewhere in our bodies and arose from them; therefore, we fear death as terrible extinction.

We live as if we were observers in a world of objects that are unchanging from moment to moment and that we perceive as a camera takes a photograph; therefore, we never really look, listen, taste, smell, touch.

We live as if our bodies, emotions, and environment obeyed mechanical laws that we can only go along with or struggle against futilely, as if there were no way we could open beyond this; therefore, there is no point in training except for survival or entertainment.

We live as if our conditioned beliefs were the only truth, our perception becomes reduced, and we feel the sacredness and the enchanted world as a threat to our sanity.

*This is the Dead World.*

Vanessa, is it surprising that so many people are in such despair? Is it any wonder that suicide rates are soaring, or that so many are taking refuge in hard drugs in a desperate search for the living world? The most profound longing of all humans—to feel the living reality of the world, the soul of the world—is snuffed out in us by these stories before we even have a chance to know it.

But what are we going to do, Vanessa? We're fed up to the teeth with being told to believe things that we see no reason to believe. We can't go back now and just start to *believe* in the living world. We can't put on another nice belief like putting on a bright new coat on top of our dirty old one.

The problem is that the story we have grown up with *prevents* us from seeing the living world. We do have the possibility of beginning to live in the living world, but only if we can see it, or at least feel it nearby. And we can feel it only if we're not prevented from believing that it exists.

I'm not trying to tell you that the Dead World is wrong and the living world is right. Both worlds exist side by side. They're the same world seen from a different viewpoint, and you'll see this as we go on. But the Dead World is terribly limited compared with the living world. It is as if you have both a black-and-white photo and a color photo of a scene: the black-and-white photo isn't wrong—it has it's own beauty and a special quality of clarity, but it leaves out a whole dimension of color.

But we're all brought up and conditioned in the Dead World, so whatever we may believe, that is the world we mostly live in. To feel the presence of the living world, we have to do a little work, to change our deep feeling toward our world. It's at this deep level of feeling that perception is conditioned so that you experience the world as dead.

To change your deep feeling, two things would be helpful. First you can practice a training that enables your awareness to penetrate deeper and deeper levels of your body-mind-feeling. For example, there are a lot of variations of training like the mindfulness-awareness practice, or sitting practice, that we do. Through such a practice, you become aware of how your body-mind is conditioned and begin to change that conditioning. A practice like this is essential, but it is not enough. If you practice without examining your beliefs in the Dead World, you may only *strengthen* your belief in the Dead World, and I have seen this happen.

The second way you can begin to open to the living world is to actually look at the assumptions about your world that you've inherited from your schooling and the culture you have grown up in. These assumptions are the story that I just told. Look at these assumptions, examine them, question them, open yourself to other possibilities.

Then you can join the two. You can bring together the insights you gain from examining your assumptions with the deepening knowledge of your conditioning that will develop in your sitting practice. In this way you can educate your deeper feelings and awareness and begin to live in a new world. When they are joined, these two—sitting practice and examining your deep conditioning—are like the two wings that can carry you soaring above the Dead World and into the living world.

In the letters that follow, I'll talk about what scientists actually know and what they do not know. I'll help you to see your conditioning. And I'll show you there is *nothing* in any of the facts or theories known to scientists that *disproves* the existence of the living world. Science could just as well have convinced you to believe in the living world as in the Dead World, if scientists had chosen that path. More than that, some of what scientists are uncovering now is *more* compatible with the living world than with the Dead World. You won't yet find many scientists ready to say that. But there are a few brave ones, as we'll see.

I will try to be very clear, as I write these letters, about what is the Dead World and what is the living world, the enchanted world. At times this may give you the feeling that it is so obvious that we could simply choose to step into the living world once and for all. But in our own lives, everyone's life, both worlds are terribly mixed up. We do not move from one world to the other once and for all, even if we long to. We have to keep seeing the Dead World, and keep leaping into the living world, over and over again. We have sudden glimpses of the living world. And when we have a glimpse we can simply relax in it. But it does close over again. Then we have to leap out again, into the living world. That is our path in life. And to do that we do need to practice, over and over again.

# LETTER 4

How Our World Was Disenchanted

*Dear Vanessa,*

Today I want to write briefly about how it happened that we forgot the enchanted world. I hope you don't mind if we go through a bit of history. This is a rather huge story, and I can only touch on the main points, but I hope you can stick with me since it's important to understand. Sometimes it can be quite enlightening to try to find out *why* you believe something.

Have you ever found yourself really not liking someone? You go on disliking her, and making snotty remarks about her to your friends, until one day someone tells you what a kind person she is. This makes you think, "Now why do I dislike that person so much?" And you actually can't remember. So you go back in your personal history and try to remember what happened between you and her that makes you so dislike her. Finally you remember, and you realize how trivial and silly it was, and you realize that you don't really dislike that person at all. So you feel tremendously relieved that you don't have to carry around that feeling of disliking her any more. Well, it is similar with the beliefs of science. When you realize *why* you believe these things, you may not feel you have to believe them anymore, just as you don't need to carry around your disliking for that person.

In the sixteenth century events came together to create what we now call the scientific revolution, which developed into the "modern, sophisticated" worldview. The idea from that revolution that has the biggest negative effect on our lives now, is the idea there is no mind in the universe other than separate, individual little minds in our brains. And this is the idea, the *invention*, that scientists hold to most vehemently and religiously. Let's see how this idea came about—how mind and life were progressively taken out of the world. Then we can free ourselves to understand how mind, in the form of awareness and feeling and even thought, could *fill* the universe, how this isn't just a primitive fantasy.

In my second letter I wrote to you about the medieval world of Europe, how similar it was to the worlds of other traditional societies. I told you about the alchemical understanding of participating consciousness, of sympathetic resonances, "as above, so below."

Throughout the Middle Ages, people lived in a cosmos with the Earth at its center, neither moving nor turning. In about the year 1250, Thomas Aquinas had figured out a brilliant synthesis of church dogma and the ancient Greek theories (that had just recently been rediscovered) about nature and the motion of the stars and planets. There had been much angry debate over whether the ancient theories conflicted with church dogma; Aquinas's story resolved that conflict and remained common belief for many centuries.

Imagine and feel that Earth-centered world, Vanessa: how and why could the stars move around the fixed Earth, and the planets appear at different points in the sky, night by night? In Aquinas's story, the five visible planets, the Sun, the Moon, and the stars, all perfect and unblemished, moved around the Earth on eight invisible spheres. Beyond these eight, there was a ninth sphere, which caused all the others to turn in their daily motion around the Earth. This system of turning spheres could explain the motions of the planets across the night sky—with some significant complications that we won't go into. Finally came a tenth sphere, the dwelling place of the creator god.

The spiritual aspirations of human beings and the nature of the physical universe were satisfyingly combined in this story. The soul's journey from the Earth to the heavenly sphere beyond the world could be envisioned as a journey guided by angels through the spheres of the Moon, Sun, and stars to the tenth sphere where the Creator awaited. And there was a neat explanation of why and how the planets, the Sun, the Moon, and the stars moved around the Earth: they were pushed around by powerful and majestic angels.

Let's look at how the medieval world broke apart and how the world we believe in, and therefore experience, came into being. I'll trace the role of science in this, but you must remember again it's a very complex story with many factors contributing to the rise of the modern world.

In the summer of 1347, a merchant ship from the Black Sea brought to Europe a horrifying disease known as the Black Death. In less than twenty years half the population of Europe died from the Black Death; the countryside was devastated, and a period of optimism and growing economic welfare was brought to a sudden and catastrophic end. People were left feeling terribly insecure and afraid. They felt especially at the mercy of the natural world.

So the century-and-a-half after the Black Death was ruled by fear and paranoia, but also by a new awakening to the value of the individual. The church felt its power waning and sought by more and more brutal means to reassert it's control. During this period, and well into the eighteenth century, more than a million people (some estimates range as high as five million) were burned as witches. Eighty percent of these were women. Their crimes were "heresies," which meant such people proclaimed beliefs and performed rituals not sanctioned by the church.

And these crimes were frequently completely trumped-up. The witches were often the healers of their village. The "good" witches, those whose healing potions and powers actually worked, were condemned even more vigorously than the "bad" witches or charlatans. And the priests often used exactly the same magic as the witches. So this holocaust had nothing to do with the truth, merely with power. More and more the church wanted to destroy anyone who had direct access to divine energies—mystics, healers, alchemists. The church was to be the only connection to the divine, the only salvation.

Perhaps you learned in school about Giordano Bruno, who is one of the heroes of the physics books, but who was also an alchemist. He is usually said to have been burned at the stake for declaring that the universe is infinite. Actually, it is more likely that he was burned for his belief that infinite mind *filled* the infinite universe and that humans could know this infinite mind directly.

A new middle class began to grow in the cities, people who realized that by controlling nature they could create comfort in this life and not have to wait for the afterlife that the church promised (while creating a veritable hell on earth). They were interested in accumulating economic wealth and in individual liberty. For these purposes power and control over nature were thought to be necessary. Salvation became secular and money came to be equated with success.

"Forget the philosophy and let's look at the details of how things actually function. Forget about emotion and feeling and quality. Let's be real." That's what people started saying then, and many continue to chant it to this day. But what about the messy, illogical, redundant, contradictory, intuitive, shadowy feelings that make up our life? "Let's look at what we *can* know—which means let's measure and quantify. Our only language will be the language of number, mathematics."

From its very beginning in this era, science was linked as much to commerce, individualism, and the accumulation of power and wealth as it was to the search for the true story of nature and humanity. Going along with this hunger for security and power, scientists regarded control over nature as the supreme goal of their work. This was clearly stated by one of the early founders of the new science, or natural philosophy, Francis Bacon. Bacon also proclaimed that nature should be put "on the rack" and her (for nature was a woman) secrets forced out of her by torture. This was his description of the new ideal of "experiment." And he wrote at the time of the witch-hunts.

☙

The denigration of women, a big story in itself, is very much connected with the control of nature and the separation of mind from body, you see. Women were considered to be of the earth, of nature, while men's souls came from the heavenly realm. Therefore women were evil and a temptation to men. Women were the original shamans and wise ones of the pagan religions and the enchanted world. And paganism was as much a challenge to the new sciences as to the church.

So another theme in this story of how mind was lost from the universe is the body's denigration and the final complete separation of mind from body. All were inventions, Vanessa. Inventions of men who repressed and denigrated their *own* bodies and who, in the custom of the time, regarded women as closer to animals than to "man." Animals, women, and all of nature were to be "mastered" by man. Intuition and feeling were to be "mastered" by reason.

Where did this hatred of the body come from? The split between mind and body began with the Greeks several hundred years B.C.E. It continued in the religious denial of the body that goes back at least to Saint Paul—deny the body and find heaven through the mind or reason. But the split was finally sealed by René Descartes, who lived at the beginning of the seventeenth century, just before Isaac Newton was born.

The only thing of which we can be absolutely certain, Descartes believed, was that we think, know, and above all doubt. But we cannot trust our senses, we cannot trust what our body tells us. The only way to know this world is to measure it and quantify it. The world was to be acted upon—not contemplated. It was more important to do than to be. We must doubt everything that we experience directly with our senses. Descartes glorified doubt into a supreme method.

From this position of doubt, Descartes paved the way for people to study the universe as if it had no mind in it. He, too, wanted to "make ourselves masters and possessors of nature." He proclaimed that mind, the knowing, thinking part of us, is completely separate from the universe we perceive with our senses, the world of things. The world of things is spread out in space, he said, whereas thinking is not extended in space at all. The world was all "other"—not us. It became a mindless object separate from us.

According to Descartes, only humans, or beings higher than them on the Great Chain of Being, such as angels, have the capacity to think, feel, or know. Other beings, such as plants and animals, have no thought and therefore no mind at all. They are completely mechanical. So if we see an animal behaving as if it were excited or affectionate or in pain, this is merely a mechanical effect, and in fact the animal feels none of these things.

Just like the bodies of animals, our own bodies are mechanical. Body is part of other—we are not our bodies. Our thought is aware of our bodies but can have no influence on them. All these ideas were simply inventions of Descartes, but you can perhaps already see in them the beginning of the modern view of mind and body. Descartes had withdrawn mind entirely from the world and put it into another sphere of being. God now was still hovering outside but played no direct part in the phenomenal world.

Descartes left behind a mechanical world, the world of bodies and what we perceive with them. And all were made of matter—mechanical, and completely devoid of awareness or feeling. The world was meaningless, with no purpose, so it could now be exploited and controlled for security as well as profit. We could have power and tools to dominate the world so we would no longer be the passive spectator at her mercy.

Descartes' philosophy struck a severe blow that effectively cut off awareness and feeling from the physical world. Another huge step toward eliminating the enchanted world from our experience was the Earth's displacement from the center of the human universe. Let's see how this happened.

In 1543 the medieval world began to unravel when Copernicus, a quiet monk, published a book suggesting that the Earth moves around the

Sun. Copernicus made his proposal only because it made it easier to calculate the positions of the planets. He did not in any way wish to oppose the church, and he made it clear in his book that he did not really *believe* that Earth moved.

Then along came Galileo, who was a pugnacious, charismatic, and courageous character. Galileo heard of an instrument that had recently been invented in Holland, called a "looker," that magnified distant objects—now we call it a telescope. He constructed a telescope and looked at the Moon through it. To his astonishment he saw that the moon had craters. Then he looked at Jupiter and saw that it had little planets, or moons. This meant that the spheres beyond our Moon were not perfect, as Aquinas's story said they were.

He called all his colleagues together and said, "Look through my telescope; you will see that the Moon is not a perfect sphere and Jupiter has moons." Some of the colleagues replied, "We do not *need* to look, we *know* the Moon is perfect and cannot have craters." Others looked but refused to believe they saw craters on the Moon, saying the telescope was faulty. Galileo was imprisoned by church authorities for saying Earth moves and the Moon and planets are not perfect spheres.

It's ironic that, in spite of Galileo's courage in going against the conventional view, the modern scientists that follow in his footsteps are just as narrow as his colleagues were. When phenomena are pointed out to them that do not fit into *their* conventional view—the view of "modern" science—they too say, "We don't need to look, we know it can't be true." People are not put in prison for going against the authority of today—the priests of science—but they are sometimes locked up in mental hospitals, which amounts to the same thing.

However, once Galileo had convinced people that Thomas Aquinas's explanation of heavenly motion was wrong, they needed to find a new explanation for the motion of the planets. And, lo and behold, the famous Isaac Newton came up with one. The Moon doesn't need anything to push it around the Earth, he pointed out. If it were left alone without anything pushing it and no friction slowing it down it would just keep on going, and fly off into outer space. So what the Moon needs is some force to keep it from flying off by pulling it in toward the Earth.

Newton's brilliant stroke was to realize that the force that pulls the Moon in toward the Earth is the same force that pulls an apple or any other object to the Earth. He called this force "gravity." He calculated that the progression of the planets around the Sun could be predicted if they, also, are kept in the Sun's orbit by this force of attraction, gravity. No more need for angels to push the planets, thank you!

With Newton's discovery of gravity, the excitement of discovering laws of nature caught on. People thought Newton had explained the motion of the planets around the Sun without needing to mention anything like mind, feeling, or soul in the world. They sought laws to explain everything in this way, including human behavior.

Eventually people lost their belief, not only in spirits that push the planets around, but in *all* spirits, all correspondences, all resonances. And since they stopped believing in them, they stopped seeing them. "As above, so below" changed to "above is above and below is below, and ne'er the twain shall meet." And since control over nature was the name of the game, what is "above"—which in Latin is *superstitio,* literally meaning "standing over"—became superstition. And, as you know, these days *superstition* is a derogatory term meaning "nonsense."

Newton himself regarded his explanation of planetary motion as a relatively small achievement and continued to study *alchemy* for the remainder of his life—a fact that is rarely mentioned in school textbooks. He considered alchemy so much more important than his theory of gravitation that he refused to publish any of his alchemical works, because he thought them too dangerous for the general population (those who could read, that is, which was only a tiny select few in those days). He did, of course, allow his work on gravity to be published, thinking this less important or dangerous.

It's interesting that the idea of "madness"—not fitting in—was also invented around this time. Madness is seeing resemblances that don't exist, as they said. Madness is being "out of one's mind"—beyond the boundary of skin and into the environment, so there's no separation. In other cultures, there has been a place for what we call madness. Strange behavior, "hallucinations," have often been a mark of developing spiritual and shamanic power. But during this period, madness was invented as another excuse to lock up people, such as visionaries or those who heard voices, whose experience went against the new, pragmatic, control-oriented philosophy. Newton, by the way, is said by some science historians to have gone mad, or to have had a nervous breakdown in his later years. And why do they say this? Because all he wanted to do was study alchemy!

❦

In this letter, Vanessa, I wanted to point out how certain men of northern Europe collectively took the turn of thought that removed mind, awareness, and feeling from their universe. And this became *our* universe, this became the modern world. This removal of mind from the universe was motivated only in small part by direct, scientific observation. The church was still all-powerful at this time—you could still be burned at the stake or tortured for expressing views that went against church doctrine. By removing mind and soul from nature, scientists were free to continue their investigations of nature without fear of the church's wrath.

It's incredible to see how the radical ideas of a few can change whole cultures, how these ideas mirror and articulate the changes already beginning to happen in society and then filter into the mainstream thinking of ordinary people. Of course, the story could be made more complex than this, with more players and many subplots. But I have given you the basic threads. When we put together these interacting threads, we have the seeds of the modern world.

Descartes, Galileo, and Newton were brilliant. Please don't misunderstand, I'm not saying that any of these people were fools or acting out of negative motives. Their clarity and insight swept away narrow, fixed beliefs based on authority and refusal to look at the world. Through their clear thinking and direct observation of nature, they gradually wrested power away from an authoritarian church from which true spirituality had long departed.

Nonetheless, following the course these men and their colleagues set, the whole world was gradually believed to be determined by predictable mechanical laws, like the law of gravity. Aquinas's spheres and intelligences were not needed because the motion of the planets had been explained by Newton's laws. And along with these, all the principles of sympathetic magic, principles of correspondences between body and nature, principles of healing touch, and so on, were discarded. They were discarded not because Newton or anyone else had proved they didn't exist, but because they didn't fit so neatly into the newly developing mechanical world that scientists were building in their imaginations.

And in the century following Newton, the idea that everything was nothing but mechanism was expanded into all areas of life—history, economics, politics, social theory, psychology, biology, physics, medicine, architecture, religion—all of it. People believed, or at least hoped, that sooner or later laws just like Newton's law of gravitation would be found to govern all of these areas of human life. This dominant way of thinking continues today in the mainstream.

From being a living organism, full and complete, the universe we live in became a lifeless hollowness, with lumps of lifeless matter in it. There was no mind anywhere in the universe of our experience, and none needed, according to this new and exciting, but terribly impoverished view of the world.

I once described to the director of a major German nuclear physics institute some results of research on precognition that I will tell you about in a later letter. These experiments seemed to me to be particularly well carried out and to give unambiguous results. I wondered aloud to this gentleman how physics might adapt to incorporate such observations. The director was a kind and friendly person who was interested in the practice of meditation, because, he said, "It follows the experimental method, just look and see." His response to my query surprised me. "There are some things we *know* are not true, and precognition is one of them. Therefore, in this case, experimental observations are irrelevant." You see, instead of letting observations affect his assumptions, he let his assumptions cancel the observations.

Even so-called "facts" are not ultimate standards of reality. What a group of scientists accept as a fact depends on the group decision of the club of scientists, and depends on what theory they currently believe in. They just *know* what to accept as a fact and what to reject as fantasy because they have been educated into that particular club. An example, again, would be any observation based on precognition. Such observations are simply not allowed into physics or neuroscience. They are not considered facts. And this is because of the shared agreement among physicists or neuroscientists that "we just *know* precognition can't happen, so we don't need to look."

Scientists are no less likely than any other humans to bring to their observations all the biases they have grown up with, all the biases we are examining in these letters. They choose the observations they make, without realizing it, to reinforce these biases. The theories that scientists formulate, the observations they make to test these theories, and the way they interpret those observations all depend on their unconscious assumptions. We are profoundly influenced by our unconscious assumptions, even if we don't know it and even if we are trained in science.

Today we regard the mechanical view as normal. Yet it's no more difficult to visualize the earth, for example, as a living organism than as a dead mechanical object we can exploit. Imposing this logical order discounts an entire world of inner reality and feeling—intuitive feeling. Harmony and trust in a sacred world *are* possible. We're not merely logical, rational, emotionless beings who can understand everything by objective analysis. Our lives are full of contradictions—we have "love-hate" relationships, we are often liberated by what frightens us. Our deepest and most meaningful experiences come from feeling and resonances of an alive and invisible depth.

The disenchantment of our world is very sad, and is why our lives often seem so drab and dead. At the same time we can't go back to the pre-technological world of the Middle Ages. We can't deny the valid, if limited, discoveries of cause and effect that science has made, or the power of logical thinking that underlies it. Rather, we have to go beyond our own dark age to rediscover the enchanted world without rejecting the positive, pragmatic aspects that science has brought to our ways of thinking and looking.

# LETTER 5

# *The Enchanted World Is* Now

*Dear Vanessa,*

Yesterday I wrote you a little bit of history. I told you how we lost the enchanted world of the medieval times. But I wonder if you are getting a bit confused now, thinking of all the other amazing phenomena that scientists have explained since the time of Newton. Perhaps by telling the stories of Galileo, Newton, and Descartes I've started you wondering, "Is there any place left in the world for the endless varieties of mind, awareness, feeling, or soul that people have felt and spoken of? Is there any place *left* for the spirits, gods, angels, kami, dralas, and all those other beings?"

Well, let me tell you something rather curious that happened to me this morning. I woke up in the middle of last night realizing that I might have ended up the letter yesterday drawing us into the same trap that our society has got into: forgetting that science is just telling stories, and *believing* the stories that seem to get rid of mind from the universe. I didn't sleep too well, wondering how I was going to get us out of the trap. When I woke up in the morning I still felt trapped.

I went to take a shower and about halfway through, when my hair was all soapy, the water suddenly turned cold with almost no warning. I quickly rinsed in cold water, dried off, dressed, and prepared breakfast. As I was doing this I was wondering how best to explain that there certainly still *is* space for the gods, or drala energies, in the world that science has created. And at the same time I was wondering why the water had run cold on this particular morning.

Then I thought, "coincidence." And I realized that "coincidence" is the answer to both questions. The answer to how the dralas fit into the real world has to do with coincidence. And, too, the water running cold was a coincidence. But at the same time it was meaningful. I could say it was a message from the dralas telling me that the answer to the problem of how they fit into the world is through meaningful coincidence. Coincidence happens at a particular moment, *now*. And scientists have absolutely nothing to say about particular moments. Science is incapable of dealing with any real particular moment.

There is, I'm sure, a physical explanation of why the hot water ran out today. Probably it had something to do with the plumbing. But scientists absolutely cannot tell why all the factors in my life came together to make the hot water run out on *that particular morning*, the morning I needed to remember coincidence. You see? There is something else involved.

How is it that a particular experience happens at *this particular moment?* We are talking about our very idea of time. We believe our lives run along a universal timeline that is the same everywhere and for everyone in the universe. So let's take a moment to look at this "time" that so dominates our lives.

To be able to write his famous laws of motion that told how a lump of matter would move when it was pushed by a force, and how the planets would move around the Sun, Newton had to imagine a couple of things. He had to imagine that there was a fixed background to all this motion; he had to be able to say that there was always the same fixed, unmoving background, anywhere in the universe, relative to which everything appeared to be moving. He called this *Absolute Space*. It was empty, passive, and did not interact with anything in it at all. It was like the stage on which the play of the universe is acted out.

And he had to imagine that there was a universal time that was the same for all the planets, and for anything else in the universe, but was not itself connected with anything in the universe—he called this *Absolute Time*.

Now, please remember that Newton *imagined* this space and this time. And he said so: "These are just hypotheses," he said.

But gradually people began to believe that Newton's Absolute Space and his Absolute Time were the *real* space and the *real* time of the *real* world we *really* live in. And this imaginary Absolute Space and Absolute Time is the space and time that you carry around with you unconsciously and that you experience the world through, in just the same way as you see the world through the perspective lines as I showed you in letter 1. This Absolute Space and Time is like the stage on which we (and scientists), just like Newton, imagine the world happening. On this stage, in this empty space-time box, scientists since the time of Newton have been imagining and building their model world. And they have convinced us to believe it is the *real* world of our experience.

We talk and think and organize our lives as if there were a real, absolute time. Scientists try to build their model of the world as if time were outside the events of the world. We have learned to believe that this real Absolute Time flows in a single line, without any loops, branches, or circularity, from the infinite past to the infinite future. And our lives are but a little blip in time. We don't usually think about this because it's very depressing. Yet our lives are completely dominated by this concept of absolute, objective time that is the same for everyone, the same throughout the universe. Time goes on without us, and will go on after we die.

We feel time as the background to everything we do, every moment of our life, a kind of blank, empty container we try to fit our lives into. We imagine it vaguely like a line on a sheet of paper. We divide time into years, months, days, hours, minutes, seconds. And the seconds are ticking, and our time is running out. So some of us try to rush around and pack as much as we can into time. What we pack into time is memories and thoughts and images of our life.

We don't experience what we do as we do it. We don't feel the changing of our life at each moment and the quality of that moment. But at the end of the day, or the end of the year, we look back and add up what we have "done" to find out how successful we were. When I was a child, whenever we used to do anything fun, my mother always used to say, "Well, this will be something nice to remember, won't it, dears?" And so we get more and more anxious and try to pack more and more experiences into our memory banks without really experiencing them. All this is so sad and depressing.

Yet there *is* no Absolute Time outside changing phenomena. There is nothing whatsoever in any field of science to suggest that there really is linear, universal, objective time. Science has simply assumed that this time is real, and we have come to believe in it and let it drive our lives . . . often into a high wall of insanity. We can't imagine life without it.

If you really examine each changing moment of your experience, you will find nothing that you could call time other than change itself. Time is not separate from changing appearances. We even measure time by looking at a clock and seeing the numbers change, or the hands move around the face. Scientists might use atomic clocks, but this measure of time is still based on something changing.

There are a multitude of influences affecting the present moment. There could even be some influence from altogether *off* the straight and narrow line of Absolute Time, which, again, is imaginary, an invention. And there are phenomena that directly contradict this narrow view of time. One of these phenomena is meaningful coincidence, and another is precognition. In precognition, influences on the present moment seem to include some kind of influence from events in the future.

Precognition, although controversial in conservative scientific circles, is now well documented by serious scientific investigation. In a later letter I'll tell you about some scientific experiments that demonstrate precognition. I'll just tell you here that in 1989 some statisticians put together the results of all the experiments that have been done in precognition over the previous half century. (What they did is called meta-analysis.) They reviewed 309 studies conducted by sixty-two different investigators, in which more than fifty thousand subjects participated in nearly two million trials. To the question, "Was precognition demonstrated overall?" the answer is an emphatic yes. The odds against the results of all these experiments happening by chance are about 1 in $10^{24}$ ($10^{24}$ is 1 followed by 24 zeros.)

But for now, let's look at a few stories of precognition.

On October 21, 1966, a coal tip slid down a mountainside in the mining town of Aberfan, Wales. It buried a school, killing 128 children and 16 adults. On the evening of October 20, a woman reported having a waking dream, which she told to six other people. "First I saw an old school nestling in a valley, then a Welsh miner, then an avalanche of coal hurtling down a mountain. . . ." This dream took place two hundred miles from Aberfan. Another person, seven days before the disaster, told two friends: "I had a horrible vivid dream of a terrible disaster in a coal mining village. It was in a valley with a big building filled with children. Mountains of coal and water were rushing down the valley burying the building. The screams of the children were so vivid I screamed myself."

There were at least two other, similar documented reports before the catastrophe. The saddest case was of a mother whose little girl had told her that morning that she dreamed she was at school and it suddenly went all black. She begged her mother not to send her to school, but her mother didn't listen.

In the sixties a well-known British poet and novelist, J. B. Priestley, announced on BBC-TV that he was conducting an investigation into unusual experiences in regard to time. He invited viewers to write to him, and he received thousands of letters. He had a team of trained, skeptical (of course) investigators whose job was to eliminate fraud and obvious error and reports that could be explained in a usual way. He was left with reports that the investigation simply could not eliminate on these grounds, which were published in his book *Man and Time*. One of Priestley's reports concerned Air Marshall Sir Victor Goddard, who, while flying in mist and rain over Scotland in 1934, became lost. He saw what should have been Drem Airfield below him. But instead of the disused hangars among fields that Drem was at the time, the airfield appeared to be in working order, with blue-overalled mechanics among four yellow aircraft. Four years later, the details of Goddard's experience were exactly fulfilled: the airport was rebuilt, training aircraft were then painted yellow (instead of silver, as formerly), and blue overalls had become standard wear for flight mechanics.

Many ordinary people have precognitions like the air marshall's, but they have no way to fit them into the world we believe in. So they dismiss them or keep them very private for fear of being thought silly. If I ever mention such things in a group of people, there is often a sigh of relief. People start pouring out stories that they've never before mentioned to anyone, but that are some of the most poignant and meaningful stories of their lives. Psychiatrists hear precognition stories in situations of psychological counseling, when people are perhaps less afraid of being considered silly and are more willing to notice and speak of their dreams and fleeting images. What makes reports like these so believable is that they are often so ordinary. They occur by chance, we can't manipulate them, and they often have no particular importance.

What all of these stories show is that we have to let go of the feeling that time is outside of us. We have to let go of the feeling that the line of time is an absolute, objective container of all our experience.

Scientists create a picture of the world with time always as the background. The world they talk about is not the actual world of *this very moment, now*. It is a generic world of generic men and women, generic dogs, generic trees, doing generic things. So it can only give generic laws. It cannot speak about this particular person, Vanessa, at this particular moment, 1:23 in the afternoon, on Sunday, March 3, 1996.

Perhaps you've noticed that there is a new trend in ads on TV now: instead of using actual pictures, they're using computer graphics. So, for example, an ad for a real estate company, instead of showing pictures of actual people driving up to an actual house, shows computer simulations; or a car ad shows a computer picture of a car driving down a computer simulation of a road.

When I first saw this I noticed it had a strange effect: the computer images seemed to penetrate my mind more than real pictures. The computer images of houses and cars and people seemed somehow superreal, more real than the real ones. Why is this? I think it's because they are idealized. A real house has patchy paint, a slightly crooked roof, and so on. But a generic computer house is perfect, ideal. It is more penetrating because it is like the *ideas* we carry around with us of a house or car, and those seem more real to us than any particular actual house or car. This is like the world science is making up.

Scientists are building up a story of an idealized world, an imaginary world, a generic world. They are creating general laws about how things generally behave. And it *seems* more real than the world we actually experience. But it still is a ghostly world, a computer-graphics world. There is no *now* in it. They can never say exactly how things will behave *at this moment*. In the world model the scientists are building, you know that if clouds gather there is a certain probability that it will rain. But you don't know *exactly* when it will start to rain, if at all. You know the water could run cold in the shower one day, but you don't know *exactly* which day that will be. You know you are going to fall in love with someone, but scientists can't tell you, and never will be able to tell, *exactly* when that will be.

An extraordinary natural phenomenon occurred in April 1987, when the great Tibetan Buddhist teacher Chögyam Trungpa Rinpoche, founder of the Shambhala teachings, died in Halifax, Nova Scotia. For a few days before and after his death, dozens of huge ice blocks, tens of feet across—mini-icebergs—flowed into the Halifax harbor. They blocked the entire harbor and caused shipping to be stopped. The Halifax harbor is no small harbor—it is the world's second-largest natural harbor—so this was no small event. And it had never happened before in living memory, nor has it happened in the ten years since. But why did the ice blocks appear at the very moment a great teacher died, who had made heroic, almost superhuman efforts to establish Buddhism in the West, and in Nova Scotia particularly? *Mere* coincidence? Or *meaningful* coincidence?

Until a few decades ago, scientists did hold on to the godlike fantasy that everything, every smallest little event, could be predicted. Following the success of Newton's laws in predicting the positions of the planets, they believed that they would sooner or later be able to predict every little thing. And you find many scientists today still carry this arrogant, godlike delusion. "We don't *yet* know . . . ," they say, with the emphasis on the "yet."

In the past twenty years, however, some scientists have been studying very complicated systems, like the weather, and have realized that these systems are unpredictable *even in theory*. A system as complicated as the weather pattern around the globe is so sensitive that a tiny change in one place could have a huge effect in another. This effect was dubbed the butterfly effect by its discoverer because, he suggested, even an event as small as the flapping of a butterfly's wings in South America could, *in theory*, cause a hurricane in the North Atlantic. This is a pretty image but it shouldn't be taken too literally—there are many factors, large and small, that go into the creation of hurricanes. There is no way it could actually be narrowed down to the flapping of one butterfly's wings. But the principle remains—a very small event can have a huge effect on a large system, like the weather, so that the system is unpredictable, *even in theory*.

The real world we actually live in simply does not follow the kind of straight-line cause-and-effect laws that scientists wish it did. There are just too many factors involved in creating any real situation, and a tiny change at one place in the universe can have a huge dramatic effect at a far-distant place.

Let's take as an example this statement by an elder and healer of the Huichol tribe, don Jose Matsuwa. When he visited California for a second time, Matsuwa said:

The last time I was in your land, we did a ceremony. I chanted with my heart. And after the ceremony, a powerful rain came. Yes, we had purified ourselves at the ocean in the morning, after celebrating through the night; then the clouds began to gather, and within several hours, it was pouring rain. You should have told me sooner that you had such problems. I would have come earlier to do a ceremony in order to change the situation.

Does a statement like this contradict the laws of science? No! Because this statement is about a particular moment, a *now*. Meteorologists can give some general rules about the weather and the probability of rain on any particular day, but they can't say exactly the moment the rain will come. And if a small quivering in Brazil can alter the weather in the Atlantic Ocean, there is no reason why a ritual arising out of generations of experience could not bring rain in California.

To experience the fullness of each moment of our life, we need to feel our direct experience of *this moment*. And this very moment is lived in our bodies, so we need to join mind and body. When we can feel the flow of our immediate changing experience, our experience can have richness and depth that has been lost for generations. When we don't try to squeeze our experience into the straight and narrow tube of objective time, we can begin to feel the large pulsating, multilayered quality of the changing patterns of our experience. We can begin to feel the way real lived time has rhythms, qualities, and even discontinuities, or gaps. At such gaps of nowness, things come together, fall together, in ways that seem uncanny. They are co-incidences, yet they carry meaning for us. And if we pay attention to them, they shock us and wake us up.

Every moment of our life is a co-incidence, which literally means "falling together"—*co-* means "together" and *incidence* comes from the Latin word for "fall." Things fall together at each moment, but science cannot and never will be able to tell us exactly which things will fall together at this very moment, *now*. And this is when resonant feelings, gods, or drala energies can enter in, only *now*.

Whatever creates a gap in your experience—stops you and brings you to the present moment—can open your heart to hear and feel the song of the dralas. And that song is often sung to the tune of meaningful coincidence. I will write more about meaningful coincidence to you in a later letter, Vanessa. For now, though, I just want to suggest that you try to notice coincidences in the coming days, as you go through your usual daily routines. You might be surprised at what you find peeping through those gaps in linear time. Sometimes a coincidence can change your life. Sometimes coincidences can bring a smile. And often they can help you to know which course to take when you have a decision to make. Always they bring meaning and richness to your life, if you don't dismiss them as "just a coincidence." Because through coincidence you can find your connection with the living world.

# LETTER 6

## First Interlude

*Dear Vanessa,*

When I was a boy, in England in the fifties, there used to be five-minute *Interludes* on TV every hour or so between programs. (And there were no commercials!) These interludes were very simple and peaceful scenes. One was of fish swimming in a fish tank. Another that I used to particularly love was of two men in a field piling wood onto a huge bonfire. They would go back and forth picking up branches and stacking them on the fire, which was taller than they were, and each time the flames would blaze up and crackle. It was relaxing to quietly watch these interludes; just to be with oneself for five minutes and have a break between the excitement of the entertainment.

In our daily lives we rarely take breaks from our busyness just to be with who we are. And perhaps this contributes to the distress and depression in our world.

In this interlude (and two more later) I will be telling you about practices that help you to see the living world and to act according to that vision. We are all used to the idea that we have to practice if we want to do well at some action, like playing the piano, painting, playing football, skiing, and so on. But you may not have thought that we also have to practice if we want to see in a new way.

For example, someone who has been blind all her life, and then has surgery that enables her to see, is not able to see the world that we do.

At first, all she sees is an incomprehensible blur of form and color. Some people going through this have become deeply depressed because they are no longer able to experience the world of sound and touch as they used to, and it takes intense training for them to be able to see the world. Or consider this description by Michael Polanyi of a medical student learning to read X-ray photographs:

> Think of a medical student attending a course in the X-ray diagnosis of pulmonary diseases. He watches in a darkened room shadowy traces on a fluorescent screen placed against a patient's chest, and hears the radiologist commenting to his assistants, in technical language, on the significant features of these shadows. At first the student is completely puzzled. For he can see in the X-ray picture of a chest only the shadows of the heart and ribs, with a few spidery blotches between them. The experts seem to be romancing about figments of their imagination; he can see nothing that they are talking about. Then as he goes on listening for a few weeks, looking carefully at ever new pictures of different cases, a tentative understanding will dawn on him; he will gradually forget about the ribs and begin to see the lungs. And eventually, if he perseveres intelligently, a rich panorama of significant details will be revealed to him: of physiological variations and pathological changes, of scars, of chronic infections and signs of acute disease. He has entered a new world. He still sees only a fraction of what the experts can see, but the pictures are definitely making sense now and so do most of the comments made on them. He is about to grasp what he is being taught; it has clicked.

Being able to see/feel the living world can only begin with knowing ourselves, and so a practice that helps us, simply and directly, to know ourselves is the first practice that I want to tell you about in this interlude. I know that you have started to practice in this way, but it won't do any harm to review it, and it might be useful for anyone else who might read these letters. Taking a break from your activity and sitting quietly with yourself is natural and absolutely necessary to be truly human. Otherwise it's very hard to come to know yourself. And doing this in a formal way is at the heart of much spiritual practice. Some call it practice of silence; others call it insight meditation or mindfulness-awareness practice.

❦

I got up at dawn today to practice meditation before breakfast, as I have each day of this writing retreat. The first practice I do is mindfulness-awareness meditation, or we sometimes call it just sitting, because that's what we do, just sit and be quiet. In this practice I sit with my legs crossed, my palms resting on my thighs. I make an effort to keep my back straight and heart area soft, and my eyes are open.

As I sit there I pay attention to my body resting on the cushion; feel the heart beating in the center of my chest and the blood pulsing through my body; experience the solidity and earthiness of my body. The practice is just to pay attention to whatever thoughts and feelings arise, as they arise, and let them go when they want to go. The effort is to be direct and honest with myself, not running away from thoughts that feel bad and not trying to hold on to good feelings. Being gentle with my thoughts and feelings, I try to be kind to them but not to encourage them unnecessarily, just to let them be there as they are.

When my attention wanders off, and I forget that I made the decision to pay attention to my thoughts and emotions, and I get *lost* in thinking something, then I bring my attention gently back to my breath. I pay attention to the breath as it goes out of my body. This gives my attention a kind of anchor that holds it here and now, as all kinds of different thoughts and emotions arise.

Sometimes this practice is very boring—I just seem to think the same thoughts over and over again and the whole thing seems pointless. Other times it's quite painful, because thoughts and feelings I've forgotten or at least not thought about for a long time come to the surface of my mind. Sometimes I get excited about a neat thought I had and then my mind goes running all over the place and I want to jump up and write. And other times it can be quite joyful. I feel present in my body and this brings warmth and clarity to my mind.

It is as if my body and my thoughts and my emotions have been running in different directions most of the time—my mind is thinking about the day ahead, my body is sleepy and slow, and my emotions are wishing for something I don't have right now (like a nice shower). But sometimes all three come together as one. For that moment I feel like a whole being. I feel as if this is how humans are meant to be—body, thoughts, and emotions playing together in harmony, like a well-tuned guitar. It is at moments like this that experiences such as the one I quoted in the first letter, by Kathleen Raine, are possible.

But it really doesn't matter whether I feel bored or upset or excited or joyful, the main point is to keep doing sitting meditation as a simple daily discipline. I feel that this is the truest way to be kind to myself—to give myself the possibility to slow down and know who I am, through and through.

It is like a naturalist looking for hours at a time, for days on end, at prairie dogs in a field, popping in and out of their holes, until he becomes so familiar with their movements that he loses his feeling of being separate from them. He begins to develop a sense of the overall patterns of behavior of the prairie dogs as well as an intimate feeling for each of the members of the tribe he's watching. At this point the naturalist may have direct insight into the behavior of prairie dogs.

In sitting practice, too, you identify with the process of your own mind until you lose the "watcher," the sense of being separate from your own mind. At that point you see directly the nature of your thoughts and emotions and the whole perceptual process. Of course, this is all easier said than done, because we have so much habitual self-deception and a tendency to avoid finding out who we are. But it can be done, if you have the courage to be honest with yourself.

The practice of mindfulness-awareness has two components, which correspond to the two aspects of the way mind functions: focusing on one activity or thought, and having a broad, panoramic sense of our experience. The mindfulness aspect of sitting practice is paying attention to our thoughts, emotions, bodily sensations. It's identifying fully with our body, thoughts, and emotions, so that there is nothing left over, no self-consciousness, no watcher, no split mind. It is not *watching* what we are doing but simply *being* what we are doing, thinking, and feeling in its smallest detail.

Mindfulness can continue when we finish sitting, when our mind, our attention, is fully present with whatever action we are executing: placing a flower, wiping a teacup, washing the car, programming a computer, anything. Our attention to detail is careful and almost deliberate.

Awareness depends on mindfulness. When we *are* fully present, openness, a broader perspective on our life, comes to us. We feel an inquisitive interest in the environment within which our actions and thoughts take place and a spaciousness and lightness in our state of mind. With awareness, we realize that our thoughts, emotions, and perceptions are not solid, heavy "things" but are simply patterns of energy.

Awareness might be experienced as a gap of openness in our solid train of thought or semiconscious chatter—a sudden flash of freshness. We cannot discover where it comes from, we cannot hold on to it, and we cannot artificially recreate it. With awareness we might suddenly glimpse a flower or see someone's face from a new perspective. It might be a touch of humor in the middle of a fit of anger.

Mindfulness and awareness are simple processes that occur naturally in every moment of our experience to a lesser or greater extent. But they are usually scattered and hidden from us and are usually unavailable to us as tools we can use. Yet people do use them instinctively.

Mindfulness is just paying attention. We all pay attention when something invites us to, for example, when we hear a strange sound or see an attractive person. You try to pay attention when you are listening to a boring school lesson, but no doubt you find your mind constantly wandering off. Even when we really want to learn a task, we often have a great deal of difficulty attending.

Awareness is opening our minds at each moment and taking in a broad spectrum of perception and feeling. Again, this can occur quite naturally. If you play a musical instrument in an orchestra, for example, you have to pay attention to the details of your own part while at the same time having a panoramic awareness of what the rest of the orchestra and the conductor are doing.

I was once talking to someone who had responsibility for training the air traffic controllers in Canada. When I told him about mindfulness-awareness, he exclaimed, "But this is just what air traffic controllers need!" He explained that these are precisely the two aspects they urgently need in their job. They need to be able to pay very precise attention to their own screen and the planes that they are directly responsible for—mindfulness. But they also need to have a constant sense of the broad picture of the airspace altogether—awareness.

Or suppose you're cooking a fairly elaborate dinner—not macaroni and cheese again! Let's say you are cooking Thanksgiving dinner. You may be cutting carrots with a sharp knife, so you need to pay careful attention to that—mindfulness. But at the same time, there is a sauce beginning to heat up, you need to baste the turkey soon, the cranberries are beginning to boil, you need to take out the wine and open it, and so on. You need to have an awareness of the whole state of the dinner, as well as careful attention to, or mindfulness of, the particular task you are doing at that moment.

You know, Vanessa, a lot of people think meditation is just exotic, Eastern, religious, navel-gazing nonsense—mostly people who have made no attempt to find out just what these practices are about. This dismissal of practices like mindfulness-awareness is probably based in the basic fear of knowing who one is.

But this practice is by no means foreign or exotic, as you know. William James, the great American psychologist who lived in the early part of this century, said, "The faculty of voluntarily bringing back a wandering attention, over and over again, is the very root of judgment, character, and will. An education which should improve this faculty would be education par excellence. But it is easier to define this ideal than to give practical instructions for bringing it about."

Well, these instructions are very practical, and they do bring it about. It's a sad reflection on the understanding of mind in our society that most people don't even know that the mind *can* be trained in this simple way. Did anyone ever tell you about this in school? No way! Isn't it pathetic!

Yet, some practice of silence like mindfulness-awareness practice is essential if we are to know our own being, our mind, body, emotions, thoughts, to *know who we are*. And a practice like this is almost essential if you are to see, and see through, the conditioning that binds you and keeps you from experiencing your ordinary world as you could, enchanted and sacred.

Well, Vanessa, now that we have had our interlude, in the following letters we will go on. In the first five letters I laid the ground by telling you the two stories, of the living world and the Dead World. I said that there are two things we must do to begin to open to the living world. We must see and unravel the conditioning that binds us to the Dead World and see the alternative possibilities, and we must have some form of direct practice to see this conditioning at work in ourselves, such as mindfulness-awareness practice. In the following letters we will journey through our conditioning and, at the same time, gradually build the story of the living world. The first step, which we'll do in the next three letters, is to look at the dance of perception that is creating our world in the meaningful coincidence of each moment.

But before we do go on, how 'bout this for a slogan:

SITTING—*a breath of fresh air!*

# LETTER 7

❦

# *Does Your Brain See?*

*Dear Vanessa,*

This morning I want to write to you about our experience of a world "out there." How do our body and mind create this experience for us? Notice I say *a* world rather than *the* world, because I'll show you that what we experience simply is not *the* world but is very much *our* version of it. And I say "create" rather than "perceive" because we usually think of perceiving as a very *uncreative* kind of thing our senses do mechanically.

There are two reasons why I want to tell you a little about this creative dance between our body-mind and the world. The first is that I want you to know how deeply the world you create from moment to moment is conditioned by the ideas that you absorbed as you grew up. I want you to know about this before we start to look more closely at the ideas that made the Dead World. If you see how you are conditioned by the ideas of the Dead World, it will be possible for you to free yourself from them, if you want to.

Later, I will write more about how you can make all this completely personal: you can actually experience the world-making process of your body-mind and perhaps change it. One way you can do this is through the sitting, or mindfulness-awareness practice that I wrote about in my previous letter.

The second reason I want to tell you about this creative dance of perception is that I want you to be able to look at your world in a new, magical way. I want you to know that, at every moment of your life, your own body-mind is dancing with the world and creating something completely fresh. Because this creative process is always fresh and always open, you can always bring in something new. You don't have to be chained to a dead narrow world by the ideas you have grown up with.

Most people don't think at all about how it is that they experience anything. But if you were to ask them, for example, "How do you see?" They would probably say that the eyes are kind of like lenses on a video camera and they take moving pictures of the world that are fed into the brain. In the brain these pictures are processed into our personal experience of the world, like a computer processes information. Most people simply assume that there *is* a world of things—trees, cars, people, and other lumps of matter—"out there." Somehow the brain reproduces a more or less accurate picture of that world, and that is what we experience as if we were watching a movie. This is called the representational theory of perception, because it says that the world is represented or pictured to you by your senses. And this just ain't so.

We actually become conscious of only a tiny fraction of the information that reaches our brain from the senses. For example, a rod (a single cell) in the retina can detect a candle flame seventeen miles away. The hair cells of the ear can detect vibrations smaller than those caused by the flow of blood through the blood vessels of the ear, and can even detect molecules bumping against the eardrum. And receptors in the nose respond to the presence of as little as four molecules of odor. Actually, this selective awareness is necessary for us to live orderly lives. If our senses did not select, restrain, and organize the barrage of messages coming to us from "out there," our life would be very difficult to manage. But perhaps one mark of becoming less brainwashed by our conditioning is that we have more energy to process the vast amount of information available to us, and can respond to it more appropriately.

Now let's look at just what the brain *does* do. There has been some fascinating work in the past ten years on the role of the brain in the way our body-mind creates its world. And to understand ourselves as fully as we can, it's very helpful to know at least a little about this work. Over the past ten years or so, brain scientists have been able to make enormously detailed maps of what areas of the brain are involved when we are thinking, experiencing particular emotions, or doing particular tasks.

To see how the brain is involved in perception-creation of our world, I want to describe, very briefly, the pathway of a visual image as it works its way through your brain. The main thing that I want you to get out of this is that the final image the brain makes available to your awareness is tremendously mixed up with your emotions, what you want to see and what you don't want to see, your preconceived ideas about the world, what you expect to see, and so on. This mixing up of your ideas with the message coming from outside you begins at the very moment the light hits your eye.

Suppose you are looking at the face of your friend Margaret. What happens after the light from Margaret's face goes through the lenses of both your eyes and makes tiny images on your retinas?

First of all, the image on the retina is upside down, reversed right to left, and broken up into over a hundred million individual electrical responses of over a hundred million neurons (nerve cells) on each retina. Different neurons on the retina respond differently to various characteristics of the image—color, intensity, form, depth, movement, and so on—and they are already beginning to break up the image into component parts. These millions of separate electrical messages then go through a tremendously complex processing in the brain. And you have a conscious perception of Margaret's face. How was your awareness of her face created as the hundred million electrical currents from each retina journeyed through the pathways of your brain?

The brain consists of several quite distinct parts. I will mention just a few of them: the cortex, the thalamus, and the limbic system (see figure 2). The cortex, the outermost layer, looks a bit like a walnut, or a piece of bark—hence the name, *cortex*, which means "bark" in Greek. The areas of the brain involved in the senses of touch and movement are arranged on top of the cortex in a pattern that is almost like an outline of the body. The cortex is divided into right and left hemispheres which control sensation and movement in opposite sides of the body—the left hemisphere controls the right arms, legs, and so on, and vice versa.

The cortex is where language, logic, interpretation, and all the "thinking" functions happen. The right and left hemispheres seem to control somewhat different ways of thinking: the left hemisphere is involved in more analytical thinking in terms of numbers and quantities, while right hemisphere thinking is more oriented to qualities and spatial arrangements—more intuitive, in other words. And language and self-consciousness seem both to involve only the left hemisphere.

Also, there are areas on the cortex where messages are received from the senses of sight, hearing, smelling, and tasting and sent back. And there are areas where interconnections *between* these senses are formed. There are millions of cross-links between all the sensory and thinking

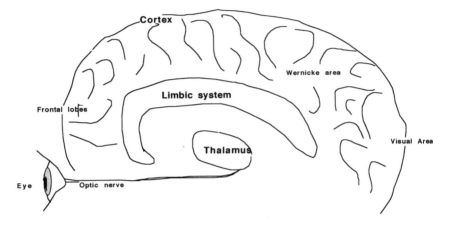

FIGURE 2. A schematic view of the brain, showing some of the areas connected to vision.

areas of the cortex, as well as millions of connections down to the lower brain.

Before they reach the cortex, the millions of fragments of the image from the retina enter an area deep in the brain called the thalamus (Greek for "inner chamber"). From there pathways carry bits of the visual image to several separate areas of the cortex. For every one fiber entering the thalamus from the retina, more than eighty times as many feed back to the same point from the cortex. So the thalamus is like a relay station, where the image of Margaret's face coming from the retina can be influenced by messages returning from the higher parts of the brain, including parts involved with emotion, thought, planning, expectation, and interpretation.

The visual areas of the cortex connect directly to the limbic system beneath the cortex—it is involved in pleasure and pain, emotions such as anger and caring, moods, and memories. The limbic system receives information from both the outside world, through the senses, and the inner world of the body. All messages from the outside have to pass through the limbic system on their way to the cortex. So moods and tones of emotion adjust the image of Margaret as it passes through the limbic system on the way to the cortex. From the main visual area of the cortex, connections also go to another area of the cortex, the frontal lobes, involved in planning for the future and for the feeling of meaningfulness.

So, altogether, there are hundreds of complicated feedback loops that connect the visual area of the cortex to all these other areas of the brain and blend emotions and thoughts with the incoming image of Margaret's face to make the final, *conscious* image.

There is one more important area of the cortex that I want to mention. It is called the Wernicke area, after the German doctor who first reported it. The Wernicke area, in the left cortex, is involved with language and the ability to form concepts or names—patients with brain damage in this area have difficulty comprehending words. But the interesting point is that this area is *also* the area used for association *between* the senses. It seems to be only in humans that connections are made *between* messages from two senses, such as eyes and touch. So it is here that the connection between the roundness and the dawn color of an orange and its sweet-tangy taste is formed. And this ability, of course, is an important part of our ability to sense whole "objects" and to give them names, in which the Wernicke area is also involved.

So, Vanessa, you can see from this simple description that the messages from the image of Margaret on your retina have to go through an incredibly complex maze of interlocking loops. These networks connect what we see with our other senses, our emotions, our interpretations, our interests, and our creation of objects with names.

And all of this feeds back to the original point where the visual information first enters the optic nerve. So your cortex is already influencing what you direct your eye to, before you even know it—a young man turns to look at a pretty woman *before* he consciously notices her. And, of course, similar systems exist for the other senses.

The whole thing seems to be going around and around in a dynamic web of loops and interconnections, a whirling pattern of electrical activity. This pattern somehow becomes what you see as Margaret's face. *But no one has the slightest idea how the electrical pattern becomes a conscious image.*

How could we ever have thought that the eye was like a camera taking its pure unadulterated little snapshots of the world "out there"? Where are we now with the idea that there is a real, solid world of "things" outside the body and that somehow the brain is able to construct an image, a "representation," of the outside world through all these complex pathways? How "true" could such an image possibly be? In the next few letters we will look more closely at how your body-mind perceives and helps to create the world you live in.

Before we leave the brain, I want to emphasize another important point here: that if a particular part of the brain is activated when someone is doing a certain action or is experiencing a certain emotion, that doesn't necessarily mean that the brain is *causing* that action or emotion. All it says is that this part of the brain lights up when you do a certain thing, just as the room lights up when you flip the light switch. That does not mean that the light switch, or the light bulb, or any of the wires between the switch and the bulb, is *causing* the room to light up, although they certainly contribute to this. We have to look deeper than that. There is also electricity, which we will not find no matter how much we chop the wires and the bulb into little pieces.

There is a powerful effort among the scientific and nonscientific authorities of our society to convince us that our mind is nothing but our brain. If this were to succeed, the mind would finally be reduced to matter—the final triumph of the Dead World.

At the same time, there are still many scientists who are not so anxious to go along with this. They speak of the brain as an instrument of the mind, or awareness, rather than as identical with the mind.

Brain scientists have done quite amazing things to find out what different parts of the brain are involved in, as we have seen—some parts are involved with processing vision, others process sound or movement, others speech, others emotions, and so on. Clearly some functions of our mind, such as the processes of vision and the other senses, are carried out by the body. But the key point is our *experience*, our *awareness of* the world.

Are your awareness, consciousness, or sense of being here, now, nothing but another function of your brain, or even of your whole body? Scientists have absolutely no idea how our *experience* comes about. Is our awareness of the sight of our friend *generated* by the brain, within the brain, or is the brain simply the instrument by which our awareness experiences that sight?

This difference between the brain as a *tool* of the mind and the brain *as* the mind is very simple, but it is very important to remember when you think or read about the brain. To help you see this, imagine you have two TVs on the table in front of you. Suppose one TV receives its power from a built-in battery, while the second TV has a solar converter and receives its power from the sun.

Now you turn on both the TVs and you see a game show, "Jeopardy," lets say. You switch channels and find a documentary on animal life, or an episode of "Seinfeld," and so on. Both TVs seem to you to work exactly the same.

Actually, there is a huge difference between the two, as we know. The first TV is generating its power, its ability to give you experiences, from within. The second TV is picking up this power from space.

The two TVs suggest the difference between awareness being the *same* as the brain (or the whole body) and the brain or body being an *instrument* of awareness. The first TV, with its built-in battery, is an analogy for the body as the *cause* of awareness, while the second TV is an analogy for the body as an *instrument* of awareness. The first TV generates its power internally, as the body might generate awareness internally. And the second TV receives its power from space, as the body might tune in to the awareness-energy-feeling of space.

I am particularly talking here of the *experiencing* aspect of what we call mind, the direct awareness or feeling of our experience. Of course, the body has a lot to do with what *kind* of world you experience, there is no doubt about that. But is it the *cause* of that experience? If awareness is generated by the body, then the first TV is a closer analogy. If awareness-feeling is all-pervasive, throughout all of space, then the body would be more like the instrument through which you, this particular person with this particular body, experience your world, just as the second TV receives its power externally, allowing it to be an instrument that enables you to experience "Seinfeld."

By the way, don't take this analogy too literally and start trying to figure out what the sun corresponds to! The point of this TV analogy is simply that it is a way of seeing that just because a part of the brain is *involved* with a particular perception, that doesn't necessarily have to mean that this part of the brain *generates* that perception.

Another way of picturing the relation between body and awareness would be to imagine awareness-feeling to be like the surface of a huge river. Each individual thing—a rock, a tree, a dog, a human, an angel, or a drala, for example—would be like a whirlpool in that river. The more complex things, like humans or dralas, create deeper whirlpools—they have more intense awareness. Everything that exists in our world is *some* whirlpool in the river of awareness, be it tiny and almost unnoticeable or huge and powerful; everything has *some* degree of awareness. And remember that our body is simply something that we perceive *within that very same awareness*. So the body is an aspect of that same awareness, just as the whirlpool is an aspect of the river. In our previous analogy, the TV would not only be picking up energy from the sunlight, it would actually be *made* of that same sunlight.

The important point that I want to get across to you here is that there is nothing, absolutely nothing, within the entire scope of commonly accepted science that actually shows whether the brain/body is the cause of our experience, our awareness, or simply the instrument of it. The decision of conventional science, that awareness is nothing but the brain, is a philosophical or religious decision. It is not a scientific decision. This decision, like any unexamined preconception, has caused great blindness in conventional science.

In later letters I'll tell you about some exciting experiments that strongly suggest the presence of awareness throughout all of space. These experiments are being done by reputable scientists with all the strictness of scientific methods. Yet conventional scientists refuse to even consider the evidence of these experiments—just like Galileo's colleagues who refused to look through his "looker."

I've started our investigation of perception in this letter with a look at the role of the brain. By looking at how the brain processes information, we've already seen how our previous ideas, our interpretations, and our emotional reactions all enter into the experience of what we see, hear, and so on. In tomorrow's letter we'll take a more inner look at perception and again see emotion, interpretation, preconceptions, and expectation combining in every moment of our experience.

# LETTER 8

❦

# *Interpretation Colors Our World*

*Dear Vanessa,*

In yesterday's letter I briefly described to you the path of vision in the brain. We saw how the message from your retina, carrying the image of Margaret's face in the form of hundreds of millions of electrical impulses, wends its way through a complex maze of neurons. It passes through parts of the brain that seem to be involved with interpretations and emotional reactions. I suggested to you that this indicates how our perception is deeply colored by our emotions and preconceptions about the world.

We don't just see a bird, we like the bird and we think it's pretty. We don't just hear a sound, we hear a sound that we dislike, like the dog barking across the lake, and call it harsh. Everything we hear, see, taste, and so on, comes with a feeling of "I like it" or "I don't like it" or "I don't really care about it." It is so close to our sense of ourself that it's very hard to see this in action. But we do need to see it and acknowledge it if we are to have any real understanding of our world.

In this letter I would like to show you another view of how what you are conscious of is already colored by your interpretations, expectations, and emotions without your even being aware of it. And there can be intelligent action based on perception without consciousness. In other words, people recognize the meaning of an object, at some level, although they don't know they are doing it.

There is a striking phenomenon known as blind sight that also illustrates that processing can take place without our being conscious of it. Some people with damage to their brain that has caused blindness—they literally cannot see anything—are able to catch a ball thrown to them. If asked to "guess" where an object is and point to it, they are able to do so with accuracy far greater than chance.

Imagine these two situations:

(1) You're sitting on a chair in the kitchen chatting with a friend. You're very absorbed in the conversation. You reach up and turn off the tap, without interrupting your train of thought. Later your friend says, "That dripping was really bothering me too." Only *then* do you become conscious of turning it off, though you were vaguely aware of doing it at the time. (We will come back later to discuss the important distinction between being conscious and being aware.)

(2) You're sitting on a chair in the kitchen chatting with a friend. You're very absorbed in the conversation. You become conscious of a noise, realize it is the tap dripping, and turn it off. You then resume your discussion.

In both of these situations, you did the same thing, turned off the tap. But the first action wasn't *conscious*.

Or have you ever had a drink of juice beside you while you were reading an engrossing book? You reach down for the glass of juice, bring it up to your lips, tip it up, and then wake up in astonishment as you find the glass is empty. You had not been conscious of drinking the juice.

Now let's look a little more closely at the processing of your world that's going on in your mind without ever reaching consciousness. Remember that in letter 7, as I was describing the pathway of a visual image through the brain, I said that it passed through the limbic system on its way to the cortex. The limbic system is the part of the brain involved in emotions, and so we would expect that visual perception would be influenced by our emotional state (and the same goes, of course, for the other senses). Here are some experiments that actually show this happening (and there are many more). They show that our perception of something is influenced by our emotional reaction to it, *without our ever being conscious of this influence.*

In one experiment, unpleasant or taboo words, such as *penis, fuck,* or *whore,* were shown to volunteers subliminally (below a level of intensity where the words could be consciously recognized). While the volunteers were watching a series of neutral or mildly pleasant pictures, these words were mixed in with the pictures but flashed too fast for them to be noticed. A taboo word produced electrical skin responses, as in a lie detector test—clear evidence that the volunteers were emotionally aroused, even though they did not consciously notice the word.

In another experiment, volunteers were shown pictures of emotionally neutral scenes, such as a boy playing a violin. For half the volunteers the picture contained the head and shoulders of a threatening and ugly man in one corner. For the other half of the volunteers the figure in the corner was a smiling face. Again, the pictures were flashed too fast to be consciously noticed. After each time they were flashed a picture, the volunteers were asked to draw and comment upon what they had seen. A significant number of the reports of the first group were distorted in a negative way; for example, they drew the central figure as dead, broken, or overlaid by a dark shadow. These results are so predictable that they were used in the selection procedure for applicants for pilots in the air forces of Norway and Sweden.

All these experiments show the effect of a subliminal emotional stimulus on *perception*. Other experiments have shown that subliminal emotional stimuli can also affect our *behavior*. In one, the word *beef* was flashed for one two-hundredth of a second (far too short for the students to notice) at seven-second intervals while a group of students were watching a film. After the showing, the students were asked to rate how hungry they were. The students to whom the word *beef* had been flashed during the movie rated themselves significantly more hungry than a group who watched the movie without the word.

Other experiments have shown that feeling often influences decision making, sometimes quite irrationally. If people are told, for example, "Imagine that you are about to buy a calculator for $15. The calculator salesman tells you that the calculator you want to buy is on sale at the other branch of the store, twenty minutes away, for $10. Would you make the trip?" Most people say that they will.

Another group is asked a similar question. This time the cost of the calculator is $125 in the original store and $120 in the branch. The majority of people presented with this version of the question said that they would not make the extra trip. In both cases the saving is the same, $5 for a twenty-minute trip. But it is far more satisfying to save $5 on a $15 purchase than on a $125 purchase. Most people go for the satisfaction rather than the logic.

Many experiments like these make it very clear that our perception, as well as the choices we make, is influenced by emotion without our being aware of it. Now let's see how interpretation and expectation similarly influence our perception. Visual illusions give us a quite dramatic way to see our mind interpreting what we see.

Look at figure 3. Are the two horizontal lines the same length? People almost invariably answer that the top line is longer. Even when we *know* that the two lines are the same length, we still continue to see the top line as longer. Once when I was about to give a lecture on all this, I was preparing a large version of figure 3—drawing it with ruler and magic marker on a large drawing pad. I measured the lines and drew them carefully. When I had finished and I stepped back to look at it, the top line looked longer than the lower line. Although I knew this was an illusion, I was so sure that it must be longer that I actually had to measure it again to convince myself that I had drawn both lines equal length!

We are used to seeing parallel lines in the actual world that project away from us into the distance—the lane lines on a highway, or railway tracks with equal-length sleepers crossing them, are examples of this. In the latter example, the farther away you look, the more the railway lines appear to converge, and the shorter the sleepers appear to become. When we look at figure 3 our brain interprets the two vertical lines as parallel lines going away into the distance instead of converging lines on a flat sheet of paper. So it thinks the top horizontal line (like the sleepers on the railway track) should appear shorter than the bottom line. But the two horizontal lines appear the same length on the flat piece of paper, so our brain thinks the top line, which *should* appear shorter, is actually longer.

This simple illusion is actually very telling. Do you remember the picture of van Gogh's bedroom that I wrote about in letter 1? Take another look at it. Remember I said that the top picture only seemed more

FIGURE 3

real because we are so used to projecting perspective lines on everything we see. Well, this illusion is a living illustration of that.

The old woman/young woman illusion in figure 4 is an example of *ambiguity*. At first you may see only the profile of an ugly old woman. Then you see the head of a young woman. Then your mind flips back and forth between the young woman and the old woman. This flipping has nothing to do with eye movements, so some higher brain function must be flipping between two possible interpretations of the picture. The brain seems to be preprogrammed to try to resolve such ambiguities by choosing between guesses.

Another type of ambiguity, illustrated by figure 5, is one in which our interpretation depends on context. Is the central figure a 13 or a B? Depends on whether you read down or across, doesn't it? That simple black mark on the paper could look like a 13 or a B, depending on the context in which you look at it.

I had an illustration of this ambiguity in perception myself yesterday morning. As I stood up from sitting practice, just after dawn, I looked out the window across to the lake, where I saw mist rising. The temperature had been below freezing for some time, so the lake had been frozen, but for the past two days it had been quite warm. So it made sense, at least it was possible, that I was seeing mist rising from the melted snow. But it didn't look quite right. The lines of the mist were too sharp, and something about the fact that it was not moving at all seemed odd. Then I realized—it was actually the shape of strips of ice between patches of water.

But no! That didn't seem right either. My mind kept flipping back

FIGURE 4

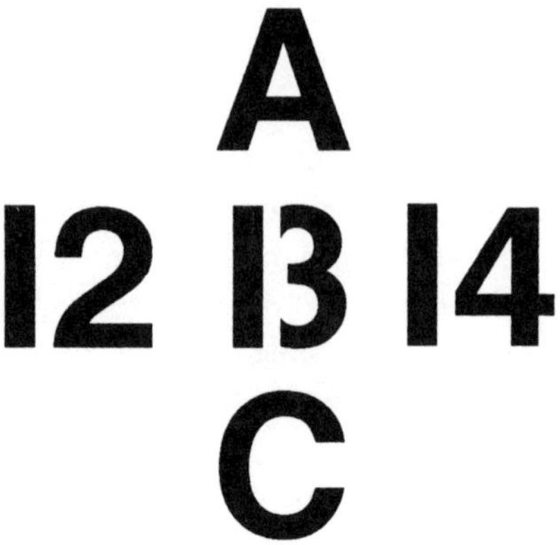

## A
## 12 13 14
## C

Figure 5

and forth between mist rising and the pattern of water and ice. I began to feel a little uneasy. Neither of these quite seemed to fit and something in my mind didn't like this ambiguity. Then my eyes glanced up to the line of pine trees on the horizon on the other side of the lake. The pattern was exactly the same as the pattern I was trying to figure out. What I had been looking at was the reflection of the line of trees in the melted ice on top of the still-frozen lake. Immediately I felt a kind of click—it now fit perfectly. And the strange thing was that from that moment on, when I glanced out the window again, I could see it *only* that way.

This morning another funny thing happened. A blue jay came chirping right outside the window and landed on the rail around the porch. I looked up and watched it, wondering what it had come for. It hopped down onto the porch and across to a flat brown piece of rock that was a few inches across and about half an inch high. The blue jay pecked at the rock a couple of times and abruptly flew away. I guess it had mistaken the rock for a piece of bread. So birds do it too!

In figure 6, do you actually see a white, solid triangle joining three points? Even when we try not to see the white triangle there, we still see it. This is sometimes called the *fiction* illusion. This kind of filling in happens in natural circumstances, as the spotted dog in figure 7 shows. Here again our brain/mind is making a guess based on seeing the dog's ears and collar. We fill in the rest of the dog based on that, even though the spots in the picture have nothing to do with a dog.

These illusions suggest that what we actually see may never be enough to tell us unambiguously what is there. So our brain-mind concludes what is "out there" in its world through a kind of tuning process. The tuning process is also influenced by what we unconsciously *expect* to find.

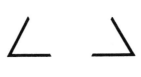

FIGURE 6

FIGURE 7

Do you remember, I mentioned in my first letter Jerome Bruner's experiments with the wrongly colored playing cards? These experiments showed that we tend to perceive something in our environment more readily if we already expect to perceive it. And we have much more difficulty than usual perceiving something that we are preset *not* to perceive. All of this applies equally to the other senses, of course.

Can you begin to see the amazingly creative dance that gives us our world? Our body-mind takes a flash of sight, a bit of sound, a soft touch, mixes these in with a large smattering of likes and dislikes, interprets them to keep us comfortably in our familiar world as much as it can, and then offers us a world. And it is a strangely circular kind of dance because that world *includes* our body-mind. Our body-mind is *part* of the world it creates and perceives. Give you a haunted feeling?

Not only does the tuning process work in our perceptions of what is outside us, it also works in trying to interpret perception of inner energies—sensations and emotions. Especially when we feel a strong emotional energy arising, we try to interpret it in familiar terms, give it a familiar name, and tell ourselves a story about it. This way we constrict the energy into a familiar emotion that we can deal with.

All these experiments suggest that as we move through the world, we don't see what is *really* there. We experience a mutual creation between what is there and the ideas and emotions that seem fitting at the time. As long as we stay in a familiar world, our perceptual system is usually able to match what is there to something familiar in our memory, more rapidly than we can consciously notice.

Just to give you another sense of how powerfully our conditioning can affect us, I will tell you about the work of psychologist Martin Seligman. Seligman suggests that personality patterns as deep as the feeling of unrelenting helplessness that some people grow up with can be a product of our conditioning. He demonstrated this dramatically and poignantly.

Seligman had three groups of dogs. The first group was given mild electric shocks that they could shut off by pushing a panel with their noses. Dogs in the second group were given exactly the same shock as the dogs in the first group, but they couldn't do anything to stop it. The third group of dogs received no shock.

After being conditioned in this classical way, the dogs were each placed in a box divided into two compartments by a low wall. When they were given a shock, they could easily jump over the wall to escape it. Now the extraordinary thing happened. When dogs from either the first or the third group were given a mild shock, they quickly learned that by jumping over the wall they could avoid it. However, when a shock was given to the dogs in the second group—those dogs who had learned that nothing they did in the first experiment made any difference—they didn't even try to jump over the wall. They simply lay down whimpering. How sad. They could have gotten out but they didn't believe they could. They had learned to be helpless. That's conditioning for you!

Seligman was able to decondition the dogs out of their learned helplessness, and this is important, you see, not just for the poor dogs' sake.

This is really the heart of what I'm trying to say in these letters. In the following letters, as we go into some more science, I'll show you the ideas that have deeply conditioned your perception and offer you alternatives to reverse that conditioning. None of this would make much sense without understanding the creative dance of perception.

Now please don't misunderstand me here. I'm not suggesting, as some alternative-healing enthusiasts did in the eighties, that every little thing that happens to you is directly caused by something you did. This moralistic idea ends up with people who have cancer believing, "It is because of something I did, and if I could only find out what it was and put it right, I would be cured." All this does is cause more pain for the people who are sick. The question is, what "you" are we talking about? Who or what is it that is dancing the creative dance of perception? It could hardly be your small little consciousness, because as we saw in the previous letter, that sometimes doesn't even enter into many of the ordinary actions you do.

So there we are, Vanessa, there is so much going on in us that we simply know nothing about. We frequently react to emotional tones without knowing why, without consciously analyzing the meaning of our perceptions. We pull back when we feel threatened, or we go toward an inviting situation, before we even realize consciously that the situation is threatening or inviting. We feel the qualities of our world and the things and people in it, and we act on those qualities often without being conscious of these feelings and responses.

The idea that we can be, or ought to be, rational beings, doing everything logically, is just another fantastic story. What we *can* do, however, is get to *know* ourselves better. We can get to know our habitual emotional responses and the way we habitually interpret the world unawares. And in this way we can learn to join our head with our heart so that we respond to our world as a whole person. The way to do this is to look directly at our experience, our thoughts, our emotions, our body sensations, and the whole process of interpreting and responding to our perceptions. And one way to do this is through the practice of mindfulness-awareness. So tomorrow I will write to you about what meditators have discovered about the process of perception.

# LETTER 9

❦

# *The Creative Dance in a Person*

*Dear Vanessa,*

Today I want to look more deeply at how meditators have described the creative dance of perception. However, before we do this I want to tell you about some fascinating clinical studies that have been done with meditators that show some of the same stages of perception that have been found by cognitive psychologists in the kind of work I described yesterday. Harvard-trained clinical psychologist Daniel Brown compared the experience of mindfulness-awareness practice in three very different traditions. He found that the three traditions describe very similar stages of meditators' experiences. He tested meditation masters and found that their experience did seem to correspond to the stages described in the classical meditation manuals.

The early stages of all three traditions correspond quite closely with the scientific understanding of the processes of perception. Meditators can become aware, for example, of the naming process in perception—how "round," "dawn colored," and "sweet-tangy" are stuck with the label "It's an orange!"

A meditator can become aware of the activity of the mind as it synthesizes messages from different senses. Back to the orange: the meditator is actually able to see the mind putting together the *visual* appearance of dawn-colored roundness with a *smell* of sweet and tangy. This is, of course, a necessary stage before the "It's an orange!" experience.

Brown found that meditators can penetrate the process of perception until they don't see a world of objects at all but simply experience the world as separate flashes of messages from different senses. Their awareness has penetrated quite deeply into the perceptual process, to experience it in action in a way that we are normally quite unconscious of. And this happens at the beginning stages of meditation!

In some other tests, lights were flashed for periods lasting just a fraction of a second, too short to be seen normally. People who had undergone an intensive three-month training in meditation were able to detect much shorter flashes than they could before they began the training, or than people could who had not been in the training course.

In a test where two lights were flashed in sequence so quickly that they normally blended into a single flash, some particularly skilled practitioners were able to detect the beginning of the flash, the flash itself, the ending of the flash, and the gap before the next flash. Flashes of light like this usually go by in a blur. This is the principle of movies, that we see continuous motion because we can't detect the change from one frame to another. Our perception is usually just too crude. But again, these experiments show that attention *can* be trained so that we see finer time intervals.

Experiencing the deconstruction of perception is not the main point of sitting practice, of course, but these studies are interesting from the point of view of scientific studies of perception. They verify that the stages of perception that have been observed by scientists, not personally in themselves but indirectly in others, in very artificial situations actually *do* take place in personal experience. It is like scientists observing prairie dogs in the laboratory and writing up what they see, and then a farmer comes along and says, "Yeah, that's the way they behave in real life too."

In the previous two letters, I have been describing perception as an observer sees it functioning in someone else. But what about perception as experienced by oneself—first-person experience? Scientists reject the possibility of observing this accurately—they generally do not know that methods, such as mindfulness-awareness training, are available to examine the deep processes of perception and awareness in direct first-person experience.

So, how *have* meditators described perception? Buddhist meditators, in particular, have seen that our experience is built up of repeating patterns. From a single moment of experience, a glimpse of your friend's face, for example, to the whole stretch of our life, our experience is a series of repeating patterns.

In a later letter I'll write about how scientists are beginning to look at some of the complex patterns in the world. They are finding that, in many situations, a large pattern is just a repetition of the same pattern on a smaller scale, which is a repetition of the same pattern on an even smaller scale, and so on. They have found repeating patterns like this in clouds, landscapes, trees, and many of the forms that appear in nature, both static forms and dynamic forms like the pattern of weather. Well, Buddhist meditators have discovered a similar repetition of patterns, from the very small to the very large, in our personal experience.

The basic pattern, the building block of the larger patterns of our life, has five parts. These are called the five *skandhas*, a Sanskrit word that literally means *constituents*. I won't go into a lot of detail about these five skandhas. But I do want to say that the Buddhists have studied the skandhas in great detail. These studies provide amazing insight into our psychology and into the process of perception.

Very simply, the five skandhas are like this. The first stage of the patterning is called *form:* in every experience we have, there is a "me" here and a "you" or "it" over there. Every experience. At every moment our senses divide the world into "me" that is here and "that" that is there. "Me," my body-mind, is here; "that," the tree, or the sound of a truck going by on the highway, is over there. That is the way all of our five senses operate: they divide the world into what they are sensing and a "me" that is doing the sensing.

We don't always divide the world according to what is inside our body, "me," and what is outside, "that." Sometimes we experience part of ourselves as "me" and another part as "that." For example, we often experience our mind as "me" and our body as "that," or our emotion as "that."

So that is a fundamental part of every experience we have. We divide our total world up that way. That is the first part of the pattern of the skandhas. And it's also the most difficult to understand; because it's so completely fundamental, we take it for granted all the time.

Then every other part follows from the first quite simply. The second skandha is *feeling:* "me" has a feeling about "that." I might like that, or I might dislike that, or maybe I just couldn't care less about that. "That" may be a person in my class. Or "that" may be a sudden sound outside my room. Or "that" may be a color or a smell. Or "that" may be my own body. Whatever it is, "me" has some feeling about "that," liking it or disliking it or couldn't care less about it.

Every aspect of the world we perceive at each moment brings with it a certain quality of feeling, positive or negative, or we simply don't bother with it in that particular moment. If you notice your own feeling as you look around your room, you can catch a sidelong glimpse of this positive-negative tone that comes with each perception. We have our favorite cup that always feels good and the favorite chair we like to sit in, some memories that always come with a nice glow. Then we have certain places that always feel hostile, certain photographs that give us a bad feeling, and so on. All of this is continually taking place at a level before all self-consciousness or naming.

Now, this is an important point about our experience that I will come back to in a moment: we don't *have* to feel something purely in terms of whether "me" likes it or dislikes it. We can feel the qualities of things *just as they are*. I can feel the blueness of the sky without having to like it or dislike it. I can feel the presence of a three-legged cat on our porch without having to like it or dislike it. Just feel the catness of the cat. But usually we don't. We feel our like and dislike much more than we actually feel the qualities of things.

The third part of the pattern of skandhas is usually called *impulse*. We have an impulsive reaction to everything in our world. And our impulsive reaction to anything, to "that," in other words, depends on how we feel about it. If we like it we want to grab it, and if we don't like it we want to get rid of it. It's as simple as that.

The fourth part of the pattern is all the *conceptual* interpretations we put on everything in our life. Every moment we are interpreting, judging, sizing up, testing, whatever we come across in our life—the people, coffeehouses, movies, music, everything. Perceptions are categorized and named according to all the complex philosophical, psychological, and practical habitual thought patterns that we carry over from past experience. Still not yet conscious, this is the level of language and concept. It consists of a vast web of associations and systems of thought: wholesome and unwholesome, religious and secular, all of our various opinions and prejudices, assumptions and preconceptions, that we subconsciously fit our experience into.

And finally, the fifth skandha, the fifth part of this repeating pattern that makes our experience, is *consciousness*. This conscious level continues in a constant stream of thoughts and conscious perceptions of all kinds. These thoughts may be thoughts about perception ("That is a red flower"); solemn, heavy, meaningful, philosophical thoughts; light, flickering, unstable thoughts; fleeting memories; hunger pangs; and so on. But they are all tied together in such a way that our feeling of a solid, safe, reasonable world of commonsense objects, meaningful relationships, and all the rest is continually maintained. Because of this stream of thoughts, which covers over and solidifies all the previous stages, we do not normally notice the process of perception in our everyday experience. We do not see the arising and ending of each moment of experience; rather, our stream of conscious thoughts produces a feeling of continuity to our experience. There are few gaps, few hints of freshness or openness.

To give you one example of how this patterning works, on a larger scale, think of the way an infant grows into the world. First, as an infant she feels no separation between her and the world. Then she begins to recognize that she has a body that is separate from other objects. She has positive or negative reactions to those objects. Later she learns to grasp and hold objects she likes and to try to avoid objects she doesn't like, even though she doesn't have any names for all these things yet. Gradually she learns names for objects and she learns what they mean to her and how to think about them. Finally, at around maybe three years old, she becomes conscious of her individuality, her identity as a personality. There you see the five skandhas. That is obviously a very simple description of how we grow into our world. But if you were to elaborate it you would find the same repeating pattern, just in more detail.

Notice that in this description of the pattern of our experience by Buddhist meditation masters, there are significant points in common with the descriptions of perception by scientists. The Buddhists, just like the scientists, see that positive and negative feeling-reactions, impulsive grasping, interpretations and judgments, all often take place without consciousness being involved.

Emotion and interpretation color and bias the way your body-mind makes your world for you, without your usually being aware of it. This may seem very obvious, when it's pointed out. But we almost always forget it in our day-to-day actions. We just assume that what we see is what is *there*. We believe and act as if we live directly in a real world. I showed you in the previous letters that this simply is not so.

I can't emphasize this too much. Because if you can actually remember this simple fact when something happens to you, or you see or hear something, then you will always be aware that *you* are there, it is *your* world. The world you perceive is not just *there*, it is a relationship *between* you and what is there. If you have disdain in your heart, you have a disdainful world; if you have affection, you have an affectionate world. If you interpret and feel the world as dead, it dies. If you feel that the world is alive, it will come to life.

You might be wondering now, "If my body-mind is so creatively involved in what seems to be 'out there,' is the world that I experience purely my own fantasy?" Why do you and I and many of our friends seem to live in the same world? Many people do think this way when they realize that the brain doesn't just take simple pictures of the world, or when they come across the cognitive experiments described in the previous letter, or when they learn about the insights of Buddhist meditators. The natural response is to say, "It's all in the mind"—by which they usually mean their little personal brain-mind.

What's happening, if you think this way, is that you have fallen off the other side of the razor's edge between two extremes. One extreme is that there is a completely real already-there world of things that our brain-mind simply *represents* to our consciousness. The other extreme is that the world we perceive is *all* our own projection: "It's all in the mind," as some people love to say.

We started this study of perception by examining the common idea that there is a real world, just like the world we perceive, that is already completely *there* before we come along to perceive it. Then we found that the brain-mind seems to play a strong role in *constructing* the world we perceive. So now we run the danger of tipping all the way over the other side of this razor's edge. We may start thinking that there is no reality at all outside of our little brain-minds—everything is nothing but a construction of our brain. It is partly because this seems so absurd that most people cling to the idea that there *must* be a real outside world.

It is very hard to see that there could be anything between these two extreme views. And the reason most people, including esteemed scientists, cannot see this is because the modern global culture is so fixated on the subject-object, inside-outside split. When people so deeply believe that the only place there could be mind is in the head, they are forced to believe that *either* there is a real world *or* it's all in the head—*either* it's outside *or* it's inside—there's no other possibility, no middle way.

If something changes in my field of vision, say a light appears suddenly in a field at night, is it outside or is it inside? If I look at the clear blue sky and see myriad little dancing lights, are the lights outside or inside my body-mind? If I suddenly hear a loud crack, like a snap of electricity, is it outside or is it inside?

Here's a strange phenomenon: Project a beam of red light onto a white wall. Hold up your hand in the path of the red light. What color would you expect the outline of your hand to be on the wall? White, of course! That's the color of the wall, and your hand is simply blocking the red light. No, what you would actually see is a blue-green hand in the middle of the red wall. I've seen it. It's quite weird. Now, where does the blue-green appearance of your hand come from? All we had was a white wall with red light shining on it. There is no blue-green light anywhere. Then you put your hand in the way of the red light, and the image of your hand on the wall is blue-green. Is the blue-green color outside or inside your body-mind? Is it *real* or isn't it? as some people would say. Shall we just dismiss it as hallucination, or shall we try to understand it?

We are so used to thinking about things from the point of view of an objective observer, the "third-person" point of view. We think about perceiving the world as if we were an external observer (like God, or a scientist in a white coat). If we think like this, then it is natural to divide the world into two: the body-mind we are observing, and its environment. But what if you imagine the situation from the point of view of your body-mind itself, from your inside? When your body-mind jumps at a loud sound, how do you determine whether it is responding to something outside or to a sudden change in its own inner state?

Are the worlds that the native shaman, or the Tibetan Buddhist yogi, or the meditation master travels to merely illusory? And are the worlds of your dreams merely illusory? Or is the ordinary waking world just as illusory, or just as real?

To someone trained to enable his or her ordinary ego-consciousness to "wake up" in dreams—lucid dreaming—the dream world seems just as real as the waking world, sometimes even more vivid. The reason we usually take the waking world to be more real is that we all have common experiences in it—we all see the same tree, or think we do. But in lucid dreaming, it is possible to enter your dream world and alter it. It is even possible for people to enter each *other's* dream worlds. There are cases where two people have been able, with training, to enter the *same* dream world. Night after night they would meet at a prearranged dream place and have adventures together, which they would confirm when they awoke. One couple eventually abandoned the experiment when the dream world they were visiting together each night started to seem more real than their waking world and they became terrified that they would not want to come back. Now which world is *really* real?

We must take a middle way between those two extremes, that the outside is reality or that it's "all in your head." The middle way is to see that, as you go through your life, your body-mind does not merely steer a way through a world already there, but it channels out a path through a range of possibilities. Rather than merely processing information coming to us from the ready-made world, we bring forth our world out of an unknown realm that is not so definite and ready-made.

Bringing forth, or enacting, our world is like a "dynamic sculpturing, or a tinkering process," says brilliant neuroscientist Francisco Varela, one of the main proponents of this middle way. He suggests, as an analogy for how our perception forges our world, the way the water from melting snow channels out rivulets as it flows down the mountainside. As the water flows down each rivulet, it makes it deeper and therefore easier for the following water to flow down—like the way so much of our perception and action becomes habitual. But there is always the possibility that the water could branch out and form a new rivulet—there is always a possibility for something completely fresh to enter our experience.

The world that our body-mind perceives, out of the totality we are immersed in, is limited by our past history and that of our environment, but within those broad limits there is tremendous flexibility about what we might experience—just like there are many possibilities for the next move of a game of chess, though it is limited by the past moves.

How come you and I experience a common world, or think we do? First, we both have human bodies. This already conditions our experience tremendously, so that your world and my world are far more similar to each other than either of our worlds is to, say, the world of a bat. We grew up in the same culture with fixed beliefs about what does and doesn't exist. Then there is language: we both recognize a pattern and give it a name, say "orange." In letter 10 we will see how our language conditions what seems to be "out there" in our world. So these three things together—the fact that we both have human bodies, we grew up in the same culture, and we speak a common language—result in there being enough in common between our two worlds that for practical purposes we assume we live in the same world.

The so-called real world is a commonly experienced fabrication out of a multitude of possible worlds. We of the modern world no longer experience other worlds, because over hundreds of years our society has conspired not to experience them. The method of science is to say that only those worlds that can be commonly experienced are real. And the scientifically based psychiatric profession takes the same approach. Any uniqueness is more and more excluded. We are becoming terrified of people who seem unique, or even eccentric, because they put into question the *reality* of our common world.

So the point is that there is not a world already there, separate from you, to be perceived passively like a camera, but *nor* is the world you perceive entirely your own creation. Rather, moment by moment, you and the world you live in are mutually molded and sculpted, as your thoughts and actions carve out the path of your living through the wholeness of which you are a part. What you perceive is limited by your conditioning, but within those limits, what you perceive and what kind of world you live in is up to you.

⁂

Before I end this letter, there is one more thing that I want to mention about the pattern of five skandhas, and that is, how they divide in our experience between mind and body. We are more used to dividing our experience up between body and mind, so the five skandhas must somehow fit into this body-mind division. Well, they do. I'll just say how, but I'm not going to go into it in a lot of detail right now.

The first skandha, how our senses divide up our experience into a "me" here and a "that" over there, corresponds to the bodily level of our being. The third through the fifth skandhas are the mental level: we become attached to things and grasp them, we interpret everything that happens to us, and we become conscious. And the second skandha, which is *feeling*, joins body and mind. I will come back to this in letter 11, when I write about how joining body and mind opens us to a new level of feeling.

Now, the chief characteristic of untrained humans, according to Buddhists, is that the five parts of the pattern, the five skandhas, pretty much carry on independently of each other. We have feelings of like or dislike that are not particularly connected to what we are really seeing or hearing at the moment. We grasp and hold on to things and people in our life without really liking them. And we try to get rid of people and things that could help us. We interpret our experience without really paying attention to our feelings. And so on and on. And we are rarely conscious of any of this.

But the tremendously, wonderfully, fantastically good news is that we *can* connect all five parts of the pattern. We *can* become a whole, harmoniously functioning person. We *can* connect our feeling with what is actually there, take in what is good for us and let go of what is not, interpret our world accurately. And so on. But to do this, we have to become aware of the whole pattern, the whole process of the five skandhas in our experience and in our being.

The parts of the pattern itself are constantly, inherently striving to work together harmoniously. It is our lack of awareness of them that is preventing this, nothing more. When we become aware of the whole process, we don't have to do anything. Just relax!

This afternoon I'll write more to you about joining body and mind, and I will tell you about a new quality of feeling, an awakened feeling, or awakened heart, that opens when we join body and mind. Right now, though, in my present experience, body is calling, singing out loud and clear to mind, "I got up really early this morning and had a light breakfast, and now I'm *really* hungry!"

# LETTER 10

## Language and a Feeling for the World

*Dear Vanessa,*

The world we live in is infinite; at any moment there are infinitely many possibilities for what we could experience. Our perception pulls a particular moment of experience out of these infinite possibilities. And an important part of how we do this is language. So, before we go any further I want to contemplate language with you a little. What is the relation between what we say, the stories we tell about our world, and what we experience? How does language dictate what and how we experience our world? This is important for understanding how we lost the magic of our world, and how we could regain it. How did we lose the ability to feel the energy patterns of gods or dralas. By refusing to name them, by saying they did not exist, science and the church gradually disabled us from perceiving them. And if we did perceive them we weren't supposed to believe they were real. They were no longer in our language, so we could no longer see them in our world.

We usually take it for granted, without a moment's doubt, that what we say corresponds to the thing or action in our world we are talking about. If I say, "I hear the cuckoo, spring must be not far away," you believe that I actually heard something go "cu-ckoo" and that I think the weather should be changing soon. But is there really this simple relation between language and what exists?

In our culture, we think in "thing" language and work on separating the parts from the whole. At school you learn that the world is full of real things, and that *facts* are true statements about these real things. You might learn about the parts of a car, how these parts are put together to make a car, and what different kinds of cars there are. Or you might learn about flowers: their different colors and shapes; their different parts, pistils and stamens, pollen, and so on; and how bees come along and, in collecting nectar from the flower, rub pollen from another flower onto it. Or you might learn about atoms and how they combine to make substances, and how substances combine to make humans.

As you get older you find a boyfriend or a girlfriend and he or she says to you, "I love you." This arouses in you a peculiar mixture of excitement and doubt that at first you cannot put a name to and so you say, "Is that true? Do you really mean that?" He or she replies, "Yes, I really feel love." At that point you decide to call your own feeling love and you say, "I love you, too."

All of this involves the belief that the words we say fit exactly with what is in the world. We believe that every sentence we say stands for the things in the world and the relations between them—nouns point to the things, and verbs or adjectives point to the relations.

So every sentence is supposedly either true or false, depending on whether it does or does not correspond to a fact. For example, suppose you say, "That tree is green," while you point to a tree. This is true if that particular tree is, in fact, green. But we can go even further than this and say that the sentence "Trees are green" is true if, in fact, all trees are green.

These seemingly rather obvious ideas have been dignified with the name "correspondence theory of truth." The correspondence theory of truth seems to work when we are talking about whether a particular leaf is green, if we just have a general idea of green. But if we really try to distinguish between green and yellow and brown, or between a tree and a bush, we start to run into some difficulties. Imagine a row of a hundred colored disks with pure deep red at one end and bright yellow at the other and all the gradations of orange in between. Now start at the red end and ask someone, "Is this red or not?" The person will, of course, have no difficulty in answering yes. And if you start at the yellow end, he or she will have no difficulty in saying no. But at which specific disk does the answer change from yes to no? We cannot answer this, yet we have a very clear sense of the difference between each disk.

Our life is full of such gradations. When does a stream become a river? When does a child become an adult? If you say "I love you" or "I hate you," these are not just matters that can be measured on a scale of one to a hundred. But you just take for granted that these sentences refer to a "fact" in the real world: a real relationship of love or hate between two real things, I and you. Americans once took for granted that "The Russians are evil" and "The Americans are good" referred to real facts about real things (people are things in that world). And so on, for millions of other things we say, or think silently to ourselves, every day.

And the correspondence theory of truth works no better for the stories of science. Whenever a new phenomenon was discovered, some thing was named to account for it: phlogiston was the name given to the cause of fire, caloric was the name given to the thing that explains heat, ether was supposed to be the thing that light waves travel in, and so on. And scientists at the time believed that there were real things in the world that corresponded to these names that they had invented—"phlogiston," "caloric," "ether." That is, until another explanation was found and some other "thing" was named. Scientists at the time believed quite certainly that phlogiston, caloric, and ether were real things. Nowadays "fields" and "quarks" are taken to be real things, and one wonders how long that will last. Many people who devoutly believe in the reality of science's stories don't seem to get the joke!

Language creates our world, as well as being a fluid, changing, partial mirror for it. In previous letters we've looked at how we experience our world in our body. And we've seen that we do not perceive through our senses a world "out there," ready-made. Our senses interact with *something* to create a world. And language plays a significant part in this world-creating that each one of us is doing each moment, without even realizing we are doing it.

Let's take an example I have used before: What is round in shape, the color of dawn, and brings a sweet-tangy taste when I bite it? Here we have three separate perceptions, with three of our senses involved: roundness, dawn colored, and sweet-tangy-tasting. Somehow we combine these together to say "an orange." Our mind puts together a bunch of sensations into a single lump and calls it a particular name, in this case an orange. The novelist Jorge Luis Borges comments, "We touch a round form, see a glob of dawn colored light, a tickle delights our mouth, and we falsely say these three heterogeneous things are one." Borges suggested, "Why not create a word, one single word, for our simultaneous perception of cattle bells ringing in the afternoon and the sunset in the distance."

Are you beginning to see how fluid our "reality" may be, and how our language is one of the ways we fix that flow—or set it free? Our language not only fixates our world but also creates new worlds. Borges's suggestion is not so far-out. What would you call the feeling that we have when we watch a flock of geese fly south on a late autumn afternoon; or when a mother waves to her son as he leaves his home to travel abroad, perhaps never to return? It is not exactly sadness, nor exactly nostalgia or poignancy.

We have no word for this feeling. And you may not even have thought that there might be a single feeling response to both of these situations, and others like them. Yet the Japanese have a word for just this feeling— *yugen*. Yugen is a feeling of harmony combined with unfathomable depth and beauty. Now that you know the word, you can recognize the feeling, and in future you will be able to say, "I feel yugen today."

Once, when I was doing research in molecular biology at Massachussetts Institute of Technology, I was sitting in a café chatting with a fellow biology researcher about the spiritual teachings of Gurdjieff, which we had just discovered, with great excitement. I tried to explain to him an experience I had had—something about being in love as far as I can remember—and I could not quite put it into words. His response was, "If you cannot describe it to me, you didn't experience it." I was profoundly startled by my friend's comment. I knew that I had experienced that feeling, and I knew exactly how it felt—I was reliving the feeling even as I was trying to describe it to my friend—yet I could not quite find the words for it.

But this is a very important point, you see. You really must question your whole feeling about language. You must see and feel how your innate, automatic belief in the correspondence theory of language and truth is constantly distorting your experience. We believe what we say, and what we think, more than we believe what we actually perceive. In fact we are so caught up in our words a lot of the time, we really don't even notice what we are actually seeing and hearing. Remember to check this out sometime, when you are walking down the street, perhaps. Catch yourself and see!

In actual life situations, the meaning and truth of even the most simple sentence depends on the context we are using it in. That's how gossip works; it takes out of context something someone said or did and circulates it around as the truth.

The meaning of what we say depends on a lot of things beyond the actual words of the sentence: variations of pitch, loudness, duration, eye movements, head nods, facial expressions, gestures, body postures, and so on. These are known as the paralinguistic (beyond language) elements of speech. For a long time I hated to use the phone because I couldn't see the person I was talking to. Finally I realized that I had to learn a whole different way of talking and listening on the phone.

If the paralinguistic elements are artificially removed, for example, by having people speak in computer-like tones from behind screens, almost all meaning goes as well. The participants in a conversation get confused, nervous, or angry; they may lose the drift of what they are saying and become more or less incoherent, and they may stop talking altogether.

Beyond this, the meaning and truth of statements depend on the larger context: "The trees are red" may be true for a four-year-old drawing pictures; "The moon is made of green cheese" for a display of cheeses depicting the solar system in a delicatessen window; "I hate you, Daddy" momentarily when Daddy takes away the candy his daughter was eating just before dinner; "He is doing very well today" for a man dying of cancer who opened his eyes and said hello to his wife; "Space is empty" for an astronaut making sure his or her space-walking suit has a functioning oxygen system; "Time passes very fast" for two lovers, and so on.

The truth of a statement about an object in a particular situation depends on how our unconscious mind fits the object into its conceptual boxes. To take a very simple example, there is a well-known riddle that goes like this: Police are called to the scene of a crime. They find in a room a dead man who is lying on the floor, two overturned chairs, a table, and fifty-three bicycles. What happened? To solve this riddle, you have to remember that there is a famous brand of playing cards depicting bicycles on the backs of the cards. And if there are fifty-*three* cards, someone must be cheating! The reason the riddle puzzles people is that they immediately categorize the term *bicycle* in one way and do not realize other possibilities.

Or take the example of someone who is asked to bring four chairs with him when he comes over for a discussion group after dinner. He brings a dining chair, a rocker, a bean chair, and a hammock, all of which work well for the discussion group. If the chairs had been requested for a formal dinner, however, the host would not have been pleased. So the meaning of the word *chairs* is not fixed in terms of what the things themselves are like. It depends on the situation, the context, it's being used in.

The way we use words points out what we value as a culture. The Inuits in the Arctic are profoundly affected by snow, as you can imagine. We call the white stuff snow and maybe qualify it with adjectives like clean or dirty. But they have at least seventeen words for different types of snow—for hard snow, wet snow, fluffy snow, and so on.

We often use words that relate to "struggle," which suggests our culture's preoccupation with conflict and aggression. Politicians are always *fighting* for our rights; doctors *wage war* on cancer; our bodies *fight off* invading bacteria; the military *fights* for peace; we *win* (or lose) arguments. This constant reference to struggle and conflict deeply affects our thoughts and behavior. Why not talk of cherishing our rights, making peace with cancer, collaborating on a discussion?

When we come to consider languages of cultures different from our own, we can see even more clearly how our language changes our experience—not just how we experience, but what there is to experience in our world at all!

A friend of ours, Trudy Sable, did several years of study with the Mi'kmaw, the native people of Nova Scotia, with Bernie Francis, a Mi'kmaw linguist. In the Mi'kmaw language, nouns are basically verbs with endings that turn them into nouns. She says about their language, "The verb is where everything happens; it is the focus of the language. . . . This focus on the verb and the copious suffixes that can be added to it allows for extraordinary breadth and creativity of expression. It makes the language adaptable, able to forge new expressions to meet the shifting and unpredictable realities of one's life. The nature of the Mi'kmaw language itself reflects the nature of the universe as being in a continuous state of flux, ever-changing and nonstatic. It also allows for great humour."

One example of this is colors. Two things about Mi'kmaw words for colors: first of all, they are all verbs. So, for example, the Mi'kmaw word for the color black literally translates as "in the process of being black." The other thing about colors is that, except for four primary colors—red, black, white, and yellow—all colors derive from other aspects of the world. So, for example, the word for blue literally says "sky-ish color" and the word for forest green says "tree-ish color." So this suggests the tremendously relational quality of the Mi'kmaq world.

Another example of the way everything is a verb is the words that were adopted by missionaries as "God." Traditionally, there is no noun in the Mi'kmaw language for the concept of a single over-lording Creator. But there are a number of different verbs that speak of different processes, words that say "creating us," "looking after us," "watching over us," "being with us."

Finally, there are suffixes that tell whether something is animate or inanimate. And for any particular object the ending can be different depending on the circumstances! For example, while most rocks are inanimate, some special rocks are termed "grandmother" or "grandfather" rocks, and have endings that tell us they are animate. Yet it is not the rock itself alone that changes; it is also the speaker's relationship to the rock. For instance, Bernie Francis told Trudy that if a rock is shaped like a bear, it might become animate, and then that rock would be referred to as a bear and would cease to be a rock. So the experience of the rock is that of a bear, and therefore it is a bear, or bearlike, and one relates to it as a conscious being.

Trudy comments, "The world is pregnant with potentialities for rocks or other objects to become animate beings with whom one can interact and communicate. Certain objects might have particular attributes and be considered to be sources of power you might keep in your possession to increase your personal power. An underlying play of energy is at work which infuses the form with animacy and consciousness."

Perhaps you begin to get a feel from this, Vanessa, how very different the world might be for a Mi'kmaw speaker. As I was explaining this to a group of people a little while ago, I actually felt a shift in my experience. For a moment the world seemed to become more fluid, energetic, and alive.

So language is as complex as the world itself, and your language doesn't correspond to a fixed real world at all. The meaning of your words changes depending on the situation in your world at that moment. And in turn, the words you use change the situation. So your world and your words are constantly changing each other. The relation between language and the world is more like dancing partners than one mirroring the other.

And it's not only when we speak to others that language filters and molds our reality; we constantly do it to ourselves too. We are continually talking to ourselves and interpreting the world to ourselves. Most people don't have any idea that this constant inner chatter is going on. But you only have to sit and do nothing for five minutes, or less, to realize that the chatter is going on, and it is unstoppable.

This constant internal chatter seems to function much deeper than your ordinary surface mind. If you try to stop the chatter or even change it, you cannot. The chatter is bound together with your emotions. It interprets the world in terms of your emotional reactions to each event of your life. This chatter is a powerful barrier to any direct experience of the world. And one of the main purposes of these letters is to show you how this chatter makes our world, and perhaps to see through the chatter to whatever might be beyond it.

Language can bind us, if we believe firmly that what we say and think about the world *is* the world. And we are even more imprisoned if we believe that only what we can describe is a true experience. But language can free us if we get the joke—I mean, if we can move in our language and see through it to a world beyond language. Needless to say, I am writing these letters in the hope that the language we use—I to write and you to read—will not bind you further but will help to set you free.

# LETTER 11

※

# *Awakening Feeling*

*Dear Vanessa,*

I have said quite a lot in these letters so far about what people have discovered in mindfulness-awareness practice. But you might be wondering, Vanessa, what the *point* is of sitting practice or any spiritual practice (not that all practices called spiritual necessarily have the same point). It is not to become a replica of someone else, the Buddha, or Jesus, or one of the many teachers around, or your psychotherapist (I know you don't have a therapist, but some people do). The point is to be genuinely and fully who you are, and then to be able genuinely to help others.

But what does it *mean* to be genuinely who you are? Who are we? Who am I? Am I fundamentally an isolated, separate individual? Yes, we certainly feel isolated and alone at times. But couldn't my "self" also participate in a much larger sense of being?

We constantly feel a barrier between ourselves and others, between our "inside" and the "outside." And we believe, and deeply feel, that this barrier is real. That barrier, and the belief that it's real, constantly distorts how we experience our world. That barrier is sometimes called the ego. But this term is a bit confusing because it is used so differently in different circumstances. So I will call the barrier that we put between ourselves and our world the *cocoon*, which is a term that Chögyam Trungpa suggested.

In the previous three letters I wrote to you about how our body-mind dances with whatever is outside of it to create the world we live in. And I showed you how our senses and mind continually try to control what we experience—we perceive what we expect or want to perceive and usually manage to keep the rest out. In this way the dance of our body-mind, our "me," and its "outside" creates the constant sense of a familiar world. And whenever something unfamiliar penetrates our perception, we feel that it is "strange" or "scary." Anything strange disturbs us or frightens us, though it may also wake us up. This noble but futile effort of our body-mind perception to continually try to create a comfortable, familiar world is how the cocoon is maintained. In the familiar world we can feel cozy and self-contained.

Through mindfulness-awareness practice you can begin to experience directly that the cocoon is not an *absolute* barrier. The cocoon is not a real, solid thing. It is a relatively useful fabrication that your body-mind creates to enable you to function in the fantasy world of things.

But as you must surely know by now, your perception of a world outside, a world of others, is a creative dance. And the cocoon is nothing other than part of this dance. This is what I've been showing in these letters. You, and your world! These two can dance together as one, and you don't have to keep up the imaginary, but absolute, separation between them that we usually feel *all* the time.

What I am writing about here is really quite simple, not at all exotic. For example, when you are watching a movie or reading a book and one of the characters really grabs you, do you sometimes completely forget that there is a Vanessa? Do you *become* the character, feel the emotions of the character just as if you were she? Or, sometimes when you are lying on the couch with your little Lhasa apso, Sernyi, do you sometimes feel so much at one with Sernyi that you know what she is feeling? It is the same with people who love one another, whether they are sexual lovers or not—at times they feel that their awareness has become one. And love, the kind of love that could also be called empathy, or feeling-with, is the means to dissolve the cocoon, the barrier between yourself and others.

If we don't feel, personally, in ourselves, the way our biases, interpretations, and emotions influence our world-making, then we will simply stay trapped in the Dead World. Our biases, interpretations, presuppositions, emotional reactions, and so on, *make up* the cocoon, as long as we don't see them. But when we do see them, the cocoon dissolves of its own accord.

As the cocoon dissolves, a new quality, or level, of perception begins to awaken for us. This is the perception that I have called feeling in early letters. *Feeling* is not really the best word for it, but I don't have a better one. Like all the words we already have in our language, it comes with so much useless baggage.

By *feeling* I don't mean heavy-duty emotions such as rage, jealousy, panic, or cloying passion. The feeling that I'm talking about is not at all the same as emotionality. Nor is it the same as sexual feeling, although sexual passion is perhaps the closest that many people come to experiencing that quality of feeling. The level of feeling I am writing about is subtler, more tender and responsive to its world.

The awkward word *feeling-with* might be better, or even *empathy* or *compassion,* which both literally mean "feeling-with." But these words sound so exalted and we tend to associate them with saintly people, or with religion, or with something else that we don't feel very connected to.

"Feeling-insight" would be another way to say it, because what I am talking about is what happens when our feeling is joined with our intelligence, or our insight. Our insight wakes up our feeling to a whole new quality, or more like a new sense organ.

I could say "love." Genuine love is the insight and affection that knows things as they are, that feels at one with things, that truly cherishes another. With that quality of love you can love anything. It could be a piece of bark from a tree, a bug, a piece of old iron, or another human. But *love* is so wrapped up with romanticism, sentimentality, pious religiosity, sex, and superficial Hollywood emotionalism that it is hard to use the word for anything genuine anymore. Genuine love can feel raw at times and can cut through to your heart like a razor. When the sun's rays melt ice—the ice will find it quite painful.

But what I am trying to say is real and very simple and I don't want it to get too complicated by introducing new terms, or old terms that don't work anymore. So I am mostly just going to say "feeling," and occasionally I'll throw in some of the other terms just to remind us that it is not the stuck grasping kind of feeling we are talking about but the feeling that has woken up.

Look back for a moment to the earlier letter where I described the five-skandha pattern that, repeating and repeating every moment, makes up our experience. You will notice that in talking about feeling, I said, "We don't *have* to feel something purely in terms of whether 'me' likes it or dislikes it. We can feel the qualities of things just as they are. . . . But usually we don't. We feel our like and dislike much more than we actually feel the qualities of things."

The usual way of feeling is "I like" or "I don't like," "he likes me" or "she doesn't like me." We interpret our world unconsciously in terms of friends and enemies, and what is nice or what is nasty, from the point of view of the barrier, or the "me." Instead of this, when the cocoon begins to dissolve, or at least get some pinholes in it, our feeling becomes a direct response to the energy of the world as it is.

You resonate with the world. Or, we could just as well say, the world resonates in you, since there is no barrier at that moment between "you" and "the world." This resonance is feeling, or feeling-with, or feeling-insight. That is exactly what resonance is, something feels-with something else. One guitar feels-with another when they are both tuned the same and you play a note on one of them.

We could summarize that whole process of cleaning up the cocoon down to the level of feeling in one phrase: "joining mind and body." It would actually be more accurate to say, "realizing the body and mind are one."

Our collective culture deeply believes that our mind is separate from our body. Our mind is supposedly rational, capable of thought, reason, knowledge. (Our mind can also be irrational, dreamy, imaginal, but that part isn't as valued as the logical part and, in the main, isn't trained.) We get the rational bit of our mind trained sitting in hard chairs in stuffy rooms hour after hour, year after year. The only way to get through this is to ignore our body's needs. I don't need to tell you this, Vanessa.

Our body, we feel, is that necessary but often embarrassing thing that is close to the animals. It wants to eat, sleep, defecate, copulate. Our bodies are functional. We can train our bodies in the gym or running in the streets. We try to make them fit our image of beauty. We eat whatever food the nutritionists tell us is good for us, or at least does not cause cancer, even though the food may be gassed, covered with wax, or stuffed with chemicals. But we have forgotten how to feel our bodies, how to enjoy them.

So we live with two separate compartments—a mind thing and a body thing. We can go for days caught up in our concepts, our thoughts, our speed, and profoundly ignore our body. We deaden our feelings. Sooner or later our ignored body has to rebel and get sick. Then we have to pay attention to a lot of pain, which usually helps us to ignore our mind. And so it goes. Even when we take the time to exercise and eat properly, we still manage to avoid really feeling our bodies.

Actually, mind and body are already joined, already one; we don't have to join them so much as go through the barrier that separates them. Mind and *body* are one—not mind and brain, or mind and head, but mind and *body*. Mind is throughout the body. Feeling is throughout the body.

Recently a whole slew of small molecules have been found that move between the brain and the rest of the body. Discovered by a feisty scientist, Candace Pert, who is not afraid to challenge the inflexible conservatism and narrow view of the establishment, these molecules are known as neuropeptides. They are made in the brain, but there are receptor sites for them throughout the immune system. The best-known of the neuropeptides are the endorphins—*endo-* meaning "within," and *-orphin* being related to the drug morphine. These are morphinelike molecules made within the body. They are naturally made pleasure drugs.

There are sites that can take up the endorphins throughout the body. They are found not only within the immune system but they're also very highly concentrated on all the sense organs. This led Candace Pert, who was one of the first to discover the endorphins, to comment that we perceive the world through a natural opiate haze.

The upshot of all this is that it is no longer possible even for scientists to claim that mind is just in the head. At the very least mind must be seen as a functioning of the whole body. As Pert says, "The old barriers between brain and body are breaking down. . . . The brain and immune system use so many of the same molecules to communicate with each other that we're beginning to see that perhaps the brain is not simply "up here," connected by nerves to the rest of the body. It is much more of a dynamic process. . . . Your mind is in every cell of your body."

Many people, especially dancers, musicians, athletes, and practitioners of awareness meditation, already know that "your mind is in every cell of your body." The point here is that people have begun to be terribly confused about this because of the strident claims of scientists over the past few decades that "mind is nothing but the output of your computer-like brain." As we have seen, however, there is now enough evidence from within science to refute this belief and to conclude that at the very least mind is throughout the body.

This split between mind and body is also a reflection of our worldview—our deeply held beliefs. It is a split between science and religion, the material and the spiritual, the dead and the alive. This split is embedded in our nervous system, in our unconscious mind, and in our social interactions. How we respond to the people in our lives, the trees and forests and mountains on our earth, the birds and frogs, depends on these beliefs.

And our understanding of spiritual practice, as well, is conditioned by these beliefs. If we don't realize and acknowledge this conditioning, and work to overcome it, our spiritual practice could just make the split deeper. We can put on a glorious embroidered robe of spiritual discipline that hides the gaping wound. Then the wound only festers and our spiritual endeavors are poison rather than medicine. But if we recognize the wound, the split between mind and body, then spiritual practice may help to heal it.

What we want to get to is the direct experience of this split, the feeling of it in our body and mind, rather than just the intellectual knowledge of it. When we experience it directly, the split will naturally heal.

The feeling of it is there, hidden, a nameless anxiety, a deep pain of the soul. Our task is to break through the barriers to feeling. Those barriers lie, of course, in our bodies and minds. They lie in our minds as unacknowledged beliefs that govern how we think and feel about the world and how we act in it. And they lie in our bodies as tensions, in our nerves, in our muscles, in our hearts.

The word *feeling* is often used when we mean emotion but also when we mean sensation—we say "I feel happy today," and we also say "I feel too hot" or "That leather jacket feels real smooth." And it's very interesting that this one word is often used to refer to the "mind," that is, emotions, as well as to the body, that is, sensations.

This shows precisely what I'm saying here—that we already do understand the oneness of body and mind. So, the key point is joining mind and body—really joining mind and body, feeling your body in your mind, feeling your mind in your body. Feel your body, feel it from the inside, feel the inner sensation of warmth in your arm, in your chest, feel the joy in this.

When you join body and mind with an inner awareness, you feel the energy of your body, the living warmth of your body, your mind filling every pore of your body. When you realize that feeling awareness extends throughout the body, then you can go further and ask, Does feeling awareness end at the boundaries of my body? Where does body end in any case? And if feeling awareness extends this far, why not farther?

That feeling awareness that spreads beyond your body is a tremendously sensitive and very real form of perception. It is the "perception of the heart," or of heart-mind joined. It is perception of the *qualities* of things. When you see blue, you don't just see blue, you feel the blueness of blue in the energy of your body; when you think of a distant friend, you don't just imagine her face, you imagine her presence, her body-mind in the space of your mind and in the feeling of your own body.

You can begin to find feeling when you pay attention not to the *content* of your sense perceptions (sight, for example) but to the *quality* of perceiving itself. Look at something you admire, a beautiful flower for example, or someone you love. Pay attention to the visual image of that object for a while, then transfer your attention from the visual image itself to the quality of the image, the feeling of it.

Feeling responds to harmony or disharmony, discriminates between colors or sounds—blue feels different from red, the squawk of a crow feels different from the call of a loon. When our body and mind respond together to our senses, that is feeling. As a way of knowing, it is tragically neglected, however, or altogether denied in the modern world. Try this on a sunny day: go outside and look at the blue sky. Become aware of how your body feels; feel the energy in your body-mind. Now transfer your gaze to something green—the grass or a tree, for example—and feel the change of quality of energy in your body-mind.

Feeling is the perception with which we actually physically experience our interconnectedness with others, with our environment, with our world. It is the perception with which we heal ourselves and others. It is the perception through which so-called paranormal phenomena happen: remote perception, psychokinesis, or even feeling the presence of angels—all depend on empathy, feeling-with.

At the feeling level of personal experience—the feeling of quality, of value, of meaning—space can be peaceful, wild, energetic, warm, or cold, and we can feel this. When we talk with another person, we can feel the space between us and the quality of it. When we are with a group of people and someone leaves or a new person arrives, the quality of the space changes. Perhaps we can remember back to early childhood when we felt this much more strongly: this end of the road is threatening, these woods are friendly, every corner of the garden has its own special quality.

Likewise, in personal experience time has qualities that we feel. The quality of early morning is different from the quality of noon or dusk. Certain days have qualities, sometimes angry, sometimes luscious or peaceful, sometimes very rich, sometimes very speedy. The seasons have their own quality. Sometimes we can detect a change in the quality of a fleeting moment. We may even have a sense of discontinuity in time.

We could almost say that feeling is like tracks, or flight paths, or optic fibers, along which awareness circulates through the space of our world, inside and outside our body-mind. It is the web that connects us with everything else in the world.

To nourish feeling we need to do two things. We need a practice, such as mindfulness-awareness, that can undo the barrier to feeling, the barrier between mind and body, the barrier between "inside" and "outside," between "me" and "you."

The other thing we need to do is to retrain our perception by feeding it new interpretations or telling it new stories, stories of the world that include feeling. For the Dead World stories have anaesthetized feeling for many years, but we can hope they have not killed it altogether.

In the next series of letters I'm going to tell you some new stories, to feed your feeling. I'll be telling you some stories of what scientists now say about the world that are quite different from the stories you've grown up with, quite different from the stories that govern and drive our society now.

Actually, we have already dealt with one of the most important of the new stories, the one about perception. We grow up with the idea that we perceive the world like a video camera takes moving pictures. This is not so much a story of science, it is just what we grow into believing, our basic conditioning. And science, as well as the psychology of mindfulness-awareness, has shown that this is not at all how it is. According to both these stories, perception is a creative dance of world-making moment by moment. And interpretation, expectation, and emotion all enter into this world-making dance. This is a very straightforward example of how science can help us to overcome our inherent mistaken beliefs about reality.

This understanding of perception means that science is as susceptible as any other human activity to the distortions and emotional biases of perception. So please always remember, the stories that some scientists are beginning to tell are just more stories. The big difference from the old stories of science is that these stories can allow feeling and awareness back into the world. Don't start to believe that they are the final truth, though. Just let them sink in and affect your feelings and perceptions. See the world through the spectacles of these new stories. Then simply rest in your awareness and trust your own experience.

# LETTER 12

## Nature, a Glorious Game of Cooperation. Surprise!

*Dear Vanessa,*

Yesterday I wrote to you about a new quality of feeling that opens when you join body and mind and begin to see through the cocoon, the barrier between "inside" and "outside," between "me" and "you." It is the quality of feeling that we discover when we let go of the self-centered emotions, narrow beliefs, and distorted interpretations of the world that weave the cocoon.

When we let go of heavy-duty emotions and feel the quivering energy within them, we find a tenderness and soreness, almost a broken heart, a feeling of joy and sadness joined together. This quivering energy of sadness-joy is tremendously sensitive and responsive to the world. This is awakened feeling.

In Sanskrit there is a term *bodhichitta*, that conveys much the same as what I am trying to convey here. *Bodhi* literally means "awake," and *chitta* has been translated as "mind" or "heart," depending, I suppose, on the personality of the translator. In Sanskrit there's no division between heart and mind, as there is in the English language.

There is a very helpful word in Japanese as well, *kokoro*. *Kokoro* is often translated by the hyphenated word *heart-mind-spirit* to show that, in *kokoro*, heart and mind are joined and something beyond both of them, "spirit," joins in as well.

But I will stick with the word *feeling*. It is warmer and more intimate than *mind*, and not so tied up with rather confused sentimental ideas as *heart*. Also, as I wrote in the previous letter, feeling is the key point in the practice of mindfulness-awareness, where we can actually alter our habitual responses to the world. And it is the link between body and mind.

Feeling is not just an imagination, a subjective psychological thing. It is a very real physical energy. It is a fine substance that scientists have not even put a name to. And when we open our feeling organ, we radiate this substance, feeling-energy. As I said in the last letter, feeling is like the tracks that awareness circulates along as it connects us with the world—really physically connects us.

Feeling-energy is very much connected with circulation and exchange. Something goes out, and something comes back. If you look at a tree or a dog or another person with awake feeling, you feel that you are sending something to them, and at the same time you feel that you are receiving something back. Don't you feel that with Sernyi, or our cat, Peter—that when your feeling goes out to them, they send something back to you?

The whole world is actually based on this principle of circulation or exchange—it is the same as interdependence. The world is built on a constant exchange of energy, a constant giving and receiving. In human lives there is simple generosity, nothing moralistic and grand, just simple generosity—giving and receiving.

Giving and receiving maintains harmony and increases the resonance between things, or between people, or between people and things.

Peoples who are not so influenced by modern ideas about human motives understand this principle that the world depends on generosity. Not just the human world, but the whole world. Take, for example, this description of Gerald Red Elk, a Sioux elder. This is written by Roger LaBorde, who was Gerald's adopted nephew and apprentice in healing:

> Some would say that Gerald's house was cluttered; all sorts of clothes, blankets, pictures, boxes, and so on, were stacked around the room. His home didn't feel that way to me, though; it had a warmth to it and a feeling of a comfortable old chair that would be nice to climb into and just rest forever. . . .
>
> Following the adoption and naming ceremonies [in which Gerald became Roger's uncle] a big giveaway was held. Blankets, star quilts, money, food, belt buckles, and war bonnets were given away to all who attended, which was quite a few. This was my first exposure to the tremendous generosity of Gerald's family. I then knew what he did with all those things stacked in his house—he gave them away.

Or think about the Trobriand islanders, who travel hundreds of miles in canoes, in ritual voyages called *kula*. These voyages are enormously dangerous. Canoes can be blown off course, or out into the open ocean, or they can smash up and sink in storms or heavy seas. The islanders believe they can be seized by giant octopuses or chased and destroyed by live stones, eaten by witches, or lured onto islands by women who are so beautiful and strong that no man can survive their passion.

And what is the purpose of these ritual voyages? Just to exchange gifts with their hosts on another island. The gifts they exchange are bracelets and necklaces, of which the hosts already possess plenty and don't really need more. Through the *kula*, the islanders maintain their links in distant networks of partners and partners' partners along the paths that will be traveled by the bracelets and the necklaces. These networks link people from different tribes, speaking different languages, over a wide area of the western Pacific.

Or consider this story of crabs, written by a Russian biologist, Petr Kropotkin, in the year 1902. Kropotkin writes,

> I was struck with the extent of mutual assistance which these clumsy animals are capable of bestowing upon a comrade in case of need. One of them had fallen upon its back in a corner of the tank and its heavy saucepan-like carapace prevented it from returning to its natural position, the more so as there was in the corner an iron bar that rendered the task still more difficult. Its comrades came to the rescue and for one hour's time I watched how they endeavored to help their fellow-prisoner. They came two at once, pushed their friend from beneath and after strenuous efforts succeeded in lifting it upright, but then the iron bar would prevent them from achieving the work of rescue, and the crab would again fall heavily on its back. After many attempts one of the helpers would go in the depth of the tank and bring two other crabs, which would begin with fresher forces the same pushing and lifting of their helpless comrade.

Generosity and cooperation are inherent qualities, not just in humans but in all animals . . . But wait! Vanessa, I hear someone calling out. She seems to be calling in a rather angry and self-righteous tone. I wonder who it is and what she is saying.

Yes, now I can hear her clearly. It is your biology teacher, Mrs. Beattie, and she is saying, "This is nonsense up with which I will not put! We *know* that all human and animal behavior is based on self-interest, only self-interest, and nothing but self-interest."

Wow! Let's ask her how she knows. "The scientists proved it long ago," she says.

Yes, it's true that science teachers, as well as most popular science writers and magazines, as well as a large number of practicing biologists, seem to believe this. But let's take a closer look at the famous theory of evolution through the struggle for survival.

Charles Darwin's theory of evolution has become firmly rooted in modern popular culture. All the animal and plant species we see in the world today, says the theory, have evolved from fewer and less complex forms over the past two or perhaps three billion years in a process of natural selection.

A multitude of life-forms appeared on the earth without the need of a creator god that is separate, transcendent, and external to this world. The glorious, miraculous world we see around is a result of its own dance of self-creation.

This basic principle of evolution cannot be doubted now. Darwin was certainly a great observer of nature and he came upon his ideas from long hours of watching finches on the Galapagos Islands during his voyages on the sailing ship *Beagle*. And Darwin's ideas opened Western minds to the inexhaustible creativity of nature.

Darwin's theory about *how* evolution occurs was a real problem, though. For some time after he realized that basic fact of evolution, he was very puzzled about *how* this happened. Then he read a book by Thomas Malthus, an economist, called *Essay on the Principle of Population*. Malthus speculated that population would continuously outpace food supply unless it was kept down by the elimination of the poor and the inept. Darwin developed an idea from this twisted thought and expanded it to become the great explanatory principle of all of nature.

He speculated that the way evolution happens is that there is not enough food to go around, so there is competition for these limited resources. This results in a struggle for survival, everyone fighting tooth and claw. If any member of a species mutates to have a small advantage, its offspring will survive better than others. The species will be "fitter." So Darwin's theory came to be called the "survival of the fittest." This was pure speculation, remember, based on an economic theory.

Actually, the phrase "survival of the fittest" doesn't say anything at all: of *course* the fittest survive—that's what we mean by *fittest*. The point is, though, that the theory that *all* evolution occurs mainly through struggle, elimination of the weakest by the strongest, competition, and aggression, *is completely unsupported by scientific evidence.*

And nor is it true that those who survive are in any way necessarily the "best," or better than species that preceded them. This means that the whole idea of "progress" is bogus. There is no reason, biologically at least, to say that humans have progressed over apes, or over bacteria, for that matter. In fact, evolutionary biologists point out that the one-celled bacteria like the ones that live in your stomach have survived far longer than humans, through ice ages and drought and meteor showers and all the rest. So from this point of view they should be considered "better" than humans. Progress is not a biological fact.

And this rather puts a damper on a lot of New Age theories about the "evolution of human consciousness"—with humans at the top of the heap, of course, and "modern" humans at that.

In the nineteenth century, "survival of the fittest" was bloated into a theory of human behavior far beyond its scope. Thomas Huxley, for example, one of Darwin's most outspoken allies, wrote, "Among primitive men the weakest and stupidest went to the wall, while the toughest and shrewdest, those best able to cope with their circumstances, but not the best in any other sense, survived. . . . So long as the natural man increases and multiplies without restraint, so long will peace and industry not only permit, but necessitate, a struggle for existence as sharp as any that went on under the regime of war."

Or consider this comment of John D. Rockefeller Sr.: "The growth of a large business is merely a survival of the fittest. . . . It is merely the working out of a law of nature and a law of God."

These statements, made by men who were respected and prestigious at the time, are sheer nonsense. Since then, this perverted view has only increased and settled deep into our cultural and individual psyches. We often feel suspicious of people's generosity, thinking they want something from us or we'll be obligated to them. It's quite difficult to simply receive, to allow others to be generous, and if we want to be openly generous ourselves, we are often put off by others.

The popular account of evolutionary theory tells us that evolution is entirely driven by self-interest, competition, and "survival of the fittest." And many biologists will tell you this even today.

We are told without a moment's hesitation that the world of nature is essentially violent and hostile. Nature programs on TV are far more often about the competitiveness in nature than about the way animals aid each other. We are told that human nature is essentially selfish and cannot be changed, and that genuine caring for others is basically impossible. Natural selection and survival of the fittest is claimed to prove this.

The religion of "struggle for survival" has entered every aspect of our culture, starting with the way children are brought up. It is taught in every high school and undergraduate biology or general science course throughout the modern world. I really cannot overemphasize how deeply this pernicious and scientifically invalid idea is ingrained into our psyche. So people now deeply believe that they are fundamentally aggressive animals and that to be aggressive is the only way to survive in this world. Many businesses have workshops for their employees teaching them to be more aggressive, and hence successful.

Yet "survival of the fittest" is dismissed as simply *wrong* by many of today's leading evolutionary biologists. One of the great modern evolutionists, Ledyard Stebbins, says, "To what extent does natural selection depend on the outcome of violent struggles or lethal combat? The answer is, very little."

It isn't always true, even in the harshest environments, that animals and plants compete against each other. According to plant physiologist Frits Went, "In the desert, where want and hunger for water are the normal burden of all plants, we find no fierce competition for existence, with the strong crowding out the weak. On the contrary, the available possessions—space, light, water, and food—are shared and shared alike by all. If there is not enough for all to grow tall and strong, then all remain smaller."

Today many biologists believe that the idea of the *fittest*, and of the supreme importance of *competition*, is simply irrelevant. Evolution, they believe, is more like a mutual creative game. In this game, organisms and the environment *both* evolve, drifting or playing together. The continually changing environment continually opens up further possible living spaces for species to evolve into. And how any species will evolve is determined by its own internal needs, its "curiosity," and the vast richness of possibilities available for it.

As Ledyard Stebbins says, "Contemporary evolutionists often describe evolution as a series of games—rather than a series of purposeful competitions—that are played because they are inevitable." Another commentator wrote that rather than a struggle, the whole of nature seems like a "glorious romp." And we humans are a part of this glorious romp.

The deluded belief that competition, survival of the fittest, and self-interest are the main driving forces of evolution is a huge obstacle to recognizing and nourishing our capacity for positive feeling. We might contrast this with a view from someone who came to the West from a

culture that took an entirely different view of humans. This is how Chögyam Trungpa Rinpoche comments on the situation he found here:

> Coming from a tradition that stresses human goodness, it was some-
> thing of a shock for me to encounter the Western tradition of original
> sin. . . . It seems that this notion of original sin does not just pervade
> Western religious ideas. It actually seems to run throughout Western
> thought as well, especially psychological thought. Among patients,
> theoreticians, and therapists alike there seems to be great concern with
> the idea of some original mistake, which causes later suffering—a kind
> of punishment for that mistake. One finds that a sense of guilt or being
> wounded is quite pervasive. Whether or not people actually believe in
> the idea of original sin, or in God for that matter, they seem to feel
> that they have done something wrong in the past and are now being
> punished for it.

This pernicious idea that there is something fundamentally bad about
human nature was taken over from the church and perpetuated in our
society through the misinformation about Darwin's theory of evolution.
The belief that human nature, or all of nature, is based on aggression,
violence, and the "shouldering aside of the weak by the strong" is simply
wrong. And many biologists recognize now that the simpleminded idea
that evolution comes about through competition and violent struggle of
one against another is *so* simpleminded that it is embarrassing.

So, if competition won't do it for us, how *has* the vast, interwoven
network of millions of species continued to exist and flourish on earth as
long as it has? Well, if fighting against each other doesn't work, how
about working together?

Cooperation, mutual aid, caring, kindness—we experience all of these
daily in our own lives. Human society would be impossible without them.
And more than that, the whole glorious romp of nature would be impossible without cooperation and a deep sense of harmony among organisms.

Cooperation has been virtually ignored by the mainstream of biology
and by science writers. Yet ecologists say that they observe ways in which
plants and animals *avoid* competition, whenever possible, far more frequently than they observe competition. They see the tendency to support
and cooperate with each other far more often than competition. "The
urge to form partnerships, to link up in collaborative arrangements,"
wrote well-known biologist Lewis Thomas, "is perhaps the oldest, stron-

gest, and most fundamental force in nature. There are no solitary, free-living creatures, every form of life is dependent on other forms."

Organisms of different species serve each other in many ways. They provide a place to stay: some crabs live in the rectums of sea urchins. They help each other obtain food: an African bird called the honeyguide cooperates with the ratel, an animal like the badger, to find and eat bee-hives.

Sometimes the relationship between the partners is so intimate that neither one would survive without the other. An example of this is the lichen, those tough greenish or bluish patches that color rock faces or hang from old trees. Lichens were for a long time considered single or-ganisms until it was found that they are actually a cooperative arrange-ment between a fungus and an alga. The combination, lichen, looks and behaves altogether differently from both the fungus and the alga. Neither fungus nor alga alone would be able to survive on a rock face, or be able to decompose the rock and take sustenance from the minerals in it. The fungus and alga can live separately in only very limited areas, but lichens are found in deserts and rain forests all the way from Alaska to the tropics.

Cooperation *within* the same species is the rule rather than the excep-tion. Most animals live and hunt together, protect each other, and play together in groups. Remember the crabs of Kropotkin? Or take African elephants: if a female or a calf is wounded, the whole group, with the oldest female taking the lead, will try to support and help the injured one away from danger.

In many situations, animals behave toward each other as if they were kin, even if they're not related biologically at all. Often, a small number of adults join together to care for a group of young, who are not necessar-ily their own. When penguins in Alaska, for example, have to go away from base camp in large groups to hunt for fish, they leave behind a few adults to care for all the young. It's almost like nursery school.

But animals go far beyond mere cooperation. Often one animal will risk its own life to warn the group. When a group of red deer flees, for example, they form a spindle-shaped pattern, with the leader in the front and the second-in-command in the rear. If they pass into a gully, the last deer will stop and stare at any intruder until the group has disappeared, endangering its own life in doing so.

The belief that animals have no feeling or awareness, another stupidity perpetuated from the time of Descartes, has led to the disgusting prac-tices of experimentation on animals and use of animals to test perfumes and detergents. This ignorant and arrogant attitude toward nonhuman animals daily causes terrible suffering in millions of other beings. There is abundant evidence that animals feel many of the same emotions of identification with each other as humans, perhaps far more intensely, because animals cannot distance their awareness from these emotions by conceptual rationalization.

I will just give you a few examples, taken from a beautiful and moving book that has recently been published, called *When Elephants Weep*. This book is filled with numerous examples of animals expressing the whole range of feeling that we experience: fear, hope, love, friendship, grief, sadness, rage, cruelty, compassion, altruism, shame, and appreciation of beauty.

First is an example of grief at the loss of a companion:

Two Pacific "kiko" dolphins in a marine park in Hawaii, Kiko and Hoku, were devoted to each other for years, often making a point to touch one another with a fin while swimming around in their tank. When Kiko suddenly died, Hoku refused to eat. He swam slowly in circles, with his eyes clenched shut "as if he did not want to look on a world that did not contain Kiko," as trainer Karen Pryor wrote. He was given a new companion, Kolohi, who swam beside him and caressed him. Eventually he opened his eyes and ate once more. Although he became attached to Kolohi, observers felt that he never became as fond of her as he had been of Kiko.

And the second example is of simple appreciation of beauty.

One afternoon a student observing chimpanzees at the Gombe Reserve took a break and climbed to the top of a ridge to watch the sun set over Lake Tanganyika. As the student, Geza Teleki, watched, he noticed first one and then a second chimpanzee climbing toward him. The two adult males were not together and saw each other only when they reached the top of the ridge. They did not see Teleki. The apes greeted each other with pants, clasping hands, and sat down together. In silence Telecki and the chimpanzees watched the sun set and twilight fall.

Vanessa, it is estimated that there are 1.6 billion different species living on the earth. And for every species there are millions or billions of individuals. Imagine the surface of the earth and how right now it teams with so many different forms of life, from the bacteria in your stomach to the great white whales, from the giant redwood trees to the lichens hanging from them, the Lhaso apso dogs, spiders, fleas, orchids, frogs, aphids, and turtles. And don't forget the galaxies and rocks, and the drala!

Imagine the brilliant colors and patterns; the sounds of chirping, buzzing, humming and howling; the smells, pungent, sweet, putrid, biting. Not to mention the colors we cannot see with our human eyes, the smells we cannot smell, the sounds we cannot hear, and the other senses that we do not even have. Such a tremendous amount of energy is being transformed at this very moment by all these living forms—eating, communicating, copulating, hunting, and playing. Try to imagine the almost unimaginable harmony and balance that is keeping this whole wonderful self-creation alive.

When we look at the whole of the natural world as one, it is clear that its well-being is dependent on a tremendous urge toward cooperation and harmony. Of course, there is competition, but it cannot be the main driving force of life on earth, and there is little biological evidence that it is.

Can it possibly be true that we are biologically bound to be selfish? No. Biologically or psychologically, we are not bound to *either* self-centeredness *or* cooperation and caring. It is up to us to make the choice.

Back in the early seventies an anthropologist, Colin Turnbull spent several years with the BaMbuti pygmies of the area of Africa then known as the Belgian Congo. He wrote a beautiful account of these people, describing their affection and kindness for each other; their profound joy in living, even in the sadness of lost loves or death; and their love of the forest, which they regarded as their provider, protector, and goddess-mother. It is the kind of story which has now become quite familiar as we read more and more stories of traditional peoples.

Some years later, Turnbull spent two years with another group, the Ik, the Mountain People. Less than a generation (about twenty years) before he visited them, these people had been hunter-gatherers moving in a vast area between mountains and forest as their food supplies changed with the seasons. They had lived very similarly to most traditional peoples, such as the BaMbuti pygmies. But, when Turnbull stayed with them, they had been confined to a small barren area of the mountains—their traditional forest hunting ground had been turned into an animal reserve and they were no longer allowed to hunt or gather berries there. The Ik lived a life of near starvation. And their relationships with each other had degenerated to constant paranoia, fear, suspicion, and competition. They laughed at each other's distress.

In one tragic scene, Turnbull describes how a circle of adults, including the mother, watched, laughing, as a baby crawled to a fire and tried to grasp a burning stick. In another poignant moment, Turnbull had given an old woman some food and pointed her in the direction of her son's hut, where she wanted to go to die, even though she knew her son would give her no food and just laugh at her agony. She took a few steps away from Turnbull's hut and then stopped, tears streaming down her face (this was the only time Turnbull ever saw one of the Ik cry.) She said that she was weeping because she had suddenly remembered a time when her people had been kind to each other just as Turnbull had been to her. And she was only forty years old, but completely aged.

The Ik had lost all connection with their land, their homeground, and their gods. Turnbull interprets their behavior as being a result of their near-starvation conditions and concludes that fear and suspicion are more basic than kindness and caring. But other peoples, such as the Australian aborigines, live in similar conditions and have not lost connnection with basic human kindness. It seems more likely that the degeneration of human relationships and community among the Ik resulted from their losing touch with the natural world that had been their home for thousands of years.

The Ik would sit for hours, doing nothing and not speaking a word to each other, just staring out across the valley from which they had been uprooted. Turnbull interprets this constant scanning as a search for food, but could it not also have been a simple longing to return home? Finally, though, Turnbull's most urgent message is that, more and more, we too are beginning to show the same symptoms of excessive and uncaring individualism as the Ik. And only we can reverse that trend, if we choose to.

Again, I say, it's up to us to choose to reconnect or to lose touch altogether with our awake heart. It's up to each of us individually, and up to us as a society. And connecting with our awake heart is not only a personal matter of connecting our own body, mind, and heart, but has to do as well with reconnecting with the natural world.

And, of course, the reason I am bothering to write about this at all is that if we *do* choose to awaken our feeling, then there are definite practical ways to do it. This is not just pie-in-the-sky philosophy.

When we join mind and body, and awaken feeling, in the practice of sitting, there are methods by which we can nourish and strengthen that awake feeling. One of these ways is taught in the Buddhist tradition and is known as "sending-and-taking," or *tong-len* in Tibetan. I will describe this practice in my next letter, which will be our second Interlude.

The basic point of sending-and-taking is very simple: you send positive energy to others, and take in and dissolve any negativity that yourself or others may be feeling. If your body, heart, and mind are joined, and you are actually in touch with your awake feeling, then you can send positive energy-feeling out to the world, and take in some of the psychic pollution that is so thick in the world today. You can become a tree! Neat, huh?

# LETTER 13

Second Interlude

*Dear Vanessa,*

In this interlude I am going to describe two practices, derived from the Buddhist tradition, that can help you to awaken positive feeling in your body-mind-heart and radiate that feeling to your world.

The first is a practice to develop the energy of loving-kindness, called *maitri* in Sanskrit, and radiate it to others. It is best to do it after a period of sitting practice, while still on your cushion. And it is best to try it only after you have some familiarity with the basic mindfulness-awareness practice.

There are several steps to this practice. You begin by developing loving-kindness toward yourself and then you radiate it to others. First you let your mind just rest for a moment in the openness of awareness that you have developed in sitting practice. You can even not focus on the breath for a moment—just let your mind rest in openness and nowness. There are several steps to the practice:

1. Recall a situation in which you felt content, free from hostility, ill will, or stress—in which you felt well-being, or happiness. Imagine this situation as vividly and clearly as you can. Recall where you were, the people involved, what you were doing, and so on. Take a few moments to establish the scene firmly. Now turn your attention to the physical sensation, the feeling, in your body as you recall that situation. Feel that sensation, its warmth, vibration, color—however you feel it. Now give it

a name—"happiness," "well-being," "contentment"—whatever feels best to you.

Still paying attention to that feeling of bodily well-being, let the details of the situation fade. Just stay with the overall mood and the feeling of well-being. As you stay with it, allow yourself to feel it intensely, allow the feeling of well-being to increase, thinking, "May I be happy" (or experience well-being, or whatever name you have given the positive mood and feeling that you generated).

2. Think of someone who is alive today with whom you have had a good relationship, toward whom you feel kindly and who has been kind to you. (It is better at first not to choose someone with whom you have too close an attachment, like a lover or a parent. Begin with someone with whom you have a basically positive connection, whom you can quite easily wish well without getting caught up in strong emotions.) Hold the image of that person vividly in mind. Recall the feeling of happiness that you generated in step 1. You may particularly experience this feeling as focused in the center of the chest—the "heart center." Feel that you are radiating that good feeling from your heart center to the person you are thinking about, thinking, "Just as I wish happiness for myself, so may So-and-so be happy."

3. When you feel some familiarity with radiating maitri to people toward whom you feel positive, you can try radiating maitri to people to whom you feel neutral, or even to people with whom you have had negative encounters. Again, recall an image of that person, radiate well-being toward him or her, and think, "May So-and-so be happy" (or experience well-being, or whatever). However, don't force it, be gentle, and stop if you begin to feel that you are not being genuine.

4. The final phase is nondirected radiation of maitri. Again generate the feeling of happiness and well-being in yourself. Now let that feeling radiate out in front of you, behind you, to both sides of you, above and below you. Without losing the feeling of well-being in your own body, focused slightly in your heart center, just feel it extending out into the space all around you. As it radiates out, let it touch whatever beings it encounters, whether they be humans, animals, plants, or the earth itself. As you radiate well-being, think, "May all beings enjoy happiness."

Finally, just as the sitting practice of mindfulness is simply practice in order to be mindful in daily life, so the sitting practice of developing maitri is also just practice toward developing maitri in daily life. When you have practiced maitri in your sitting space and feel some familiarity with it, then you can begin to practice it in daily life as well. When you are with someone toward whom you feel kindly, and who has positive feelings toward you, you can quietly recall your feeling of well-being that you have generated in your practice and radiate that to your friend. And again you can move on to radiating maitri in more difficult situations when you feel ready.

⁊

The second practice, sending-and-taking, or tong-len, is an extension of the practice I just described, on developing maitri. When you do the maitri exercise, you may find that as you radiate kindness to others you begin to think of all the pain in the world, the suffering of others in less fortunate lands, or the suffering of a friend that is sick or perhaps you are reminded of a difficult time, a moment of sadness or anger, perhaps, in your own life. The practice of sending-and-taking helps you to work with the sadness of realizing others' pain. Usually we feel that we have to try to keep pain away from us if we are to remain happy ourselves. This practice is based on the fundamental fact that we are interconnected with each other and we cannot separate ourselves off and try not to be touched by the pain and troubles and stresses of others.

The practice is very simple: we send well-being to others, and we accept their pain into ourselves. This is done using the breath as a vehicle. Having practiced the development of maitri for a while, you can begin to radiate maitri on the out-breath. On the in-breath you open yourself to allow in the pain and sadness of the world. You allow into your being the suffering, anxiety, stress, and darkness of all those trapped in the cocoon world, and let yourself feel the sadness of all this. After your breath comes in, you simply let go of that suffering. Transfer your attention to your feeling of well-being, and on the out-breath you breathe well-being, health, benevolence, from your heart to others and the environment.

You might begin drawing in the pain of a particular incident that you recall, in which someone was physically or emotionally hurt. It could be a time when you yourself were hurt, or when you empathized with the pain of a friend. Draw in that pain on your in-breath, and then radiate out loving-kindness on your out-breath to the person who was suffering—yourself or your friend. After doing this for a while so that you really feel the texture of that particular emotional or physical pain, you can open further to acknowledge that many others feel that same kind of pain. Now imagine that, on your in-breath, you are drawing in everyone's pain like that. And again radiate loving-kindness on the out-breath.

This practice is a recognition that you really are not separate from others, that the pain of others is your pain, that you cannot actually generate well-being in yourself unless you are willing also to work with the pain of others. You cannot truly isolate yourself in that way and pretend that the sadness of the world does not affect you. So you allow the sadness of others to come into your being without putting up barriers, just as your breath naturally comes in if you do not try to stop it. And in exchange you give away well-being.

This practice can be a difficult one to do—you might experience quite a lot of resistance to letting in pain, it might make you feel quite claustrophobic. However, you do not get stuck on your sadness, because it is just for one breath, and then on the out-breath you relax and breathe out goodness again. Likewise, you cannot get stuck on your joy, because on the in-breath you allow sadness into your heart again. So this is a good practice to help you realize the inseparability of sadness and joy, just as your in-breath and your out-breath are really all part of one breath, which is part of the atmosphere that we all share. It is better, though, not to try the practice until you feel some relaxation and familiarity with basic mindfulness-awareness practice and the practice of radiating maitri.

❦

These practices are effective and quite magical in helping you to let go of your self-centeredness and open your heart to sadness and joy, and to actually care for others.

In the next letter, we are going to plunge deeper and deeper into the space of feeling-energy-awareness. But before we start, how about another slogan:

RADIATING KINDNESS—BENEFITING YOURSELF
AND OTHERS.

# LETTER 14

❦

# *Boundaries in Space: The Stuff the World Is Made Of*

*Dear Vanessa,*

In letter 11 I wrote about feeling that radiates to space when we let go of the cocoon and open our heart-mind, our feeling organ. I've suggested quite a few times in these letters that awareness fills all of space. So in the next letters I want to go into what this might mean.

First, I want to clear up some things about what I mean by *mind* and *awareness*. This is not easy to do, because both words are used so loosely in our language. Since people gave up contemplating these things at all deeply, the words have become more and more trivialized and confused. Today, people usually confuse mind with "thinking." They ignore the "mind" part of feeling, emotion, and bodily sensation.

So I want to clarify how I'm using these words, because it gets quite important to be clear about this when I say that awareness fills the whole of space. By *mind* I mean the whole kit 'n' caboodle of thinking, feeling, emotion, awareness, consciousness, everything that we usually think is "in here" in contrast to "out there." So that's a very broad term.

Now let's look at awareness. Sit quietly for a moment and be aware of what is going on in your mind. . . .

OK? What were you aware of? You may have been aware of thoughts ("Gee, this is really interesting" or "What's Dad getting at *now?*" or "I'm thinking") or emotions ("boring," "exciting," "irritating") or bodily sensations ("too hot," "chilly draft," "hard on buttocks"). There *is* something else, though. Yes? There's *awareness itself.* There is a feeling that is not *itself* thinking, or emoting, or sensing, it is just *being aware,* silently, without comment or judgment. We usually take awareness completely for granted. "There is a tree," we say. Yes, but also, "There is the awareness of a tree."

Now, awareness is not necessarily conscious. This is a very important point, although it's a bit subtle. It is crucial for understanding what I'm trying to show you in these letters about awareness. Consciousness involves watching what you're feeling or doing, rather than just feeling it or doing it. In consciousness there's always thought—you're *thinking* what you're perceiving or doing as it's happening. You don't just look at a tree, you think "tree" as you look at it. And at the same time you are conscious of your self, the "I" who's thinking or perceiving or doing. It's actually really simple, and I'm sure you get it, but I'll give you some examples to try to illustrate what I'm saying.

Remember the little scene I pictured in letter 8 of you turning off the tap? In the first instance, your "I" was not *conscious* of turning off the tap. You continued your conversation, and that was what you were conscious of. But your body-mind was *aware* of the dripping, of moving across the floor to the sink, of reaching out to the tap, and so on.

Here's another example: the great hockey player Wayne Gretsky said that when he is on the ice he has no thought, yet he is intensely aware of the position of every player on the ice rink. If someone moves, he moves; when the puck comes toward him he feigns a move one way and then moves in another because he knows that just behind him a defender is approaching; and so on. Yet he is not conscious of all those things; he is not conscious of little Wayne; and he is not thinking about what he is doing. He is just doing it. If he tried to be conscious of it all, he would be a lousy player.

Or consider what this rock climber has to say, "It's a Zen feeling like meditation or concentration. One thing you're after is the one-pointedness of mind. You can get your ego mixed up with climbing in all sorts of ways, and it isn't necessarily enlightening. But when things become automatic, it's like an egoless thing, in a way. Somehow the right thing is done without you ever thinking about it or doing anything at all. . . . It just happens."

The cocoon, little "me," makes a big fuss about consciousness. When people talk, for example, of living life more consciously, they sometimes seem to mean trying to make their *thinking* dominate their life even more. Their conscious thought wants more control over the rest of their body-mind. It's terribly tiring, and terribly harsh! But when we relax our effort to be conscious, we can tune in to awareness, which does not care who it belongs to. We have to crank up consciousness, but we don't have to crank up awareness, because it is all-pervading, whatever we do. We just have to open to it.

Scientists and philosophers have a difficult time understanding this simple but extremely significant point about our experience. The reason? Because they are so *enthralled*, almost obsessed, with thinking. People who are more in touch with their bodies, however, have less trouble understanding this. Dancers, rock climbers, hockey players—the best of them all know that at their best moments, when they are 100 percent intensely aware of what they are doing, there is no conscious thought at all.

You might be thinking, "But isn't what you are calling awareness just something in my brain, or at least my body?" Well, that is what these next few letters are about. Remember I suggested the image of the two TV sets as two possible models for the relation between mind and body? I suggested that the TV that picked up its energy from space could perhaps be a better image of awareness-energy than the TV that generated its energy inside itself. Then I suggested that mind could be like a river, and all the things in our world like whirlpools in the river. Well, now I'm narrowing it down a little more: what fills all of space is two particular aspects of mind—awareness and feeling. *Feeling* forms the pathways and webs of pathways for *awareness* to ride on.

Now, where did that come from, you might ask? Well, it's common to just about every way of talking about and experiencing the world, except the way of science. It's what people mean nowadays, I think, when they speak of soul in the world. Soul is awareness and feeling combined. And there's energy there as well. And the various nonhuman beings that people experience—deities, angels, fairies, spirit helpers, dralas—these too are patterns of feeling, energy, and awareness combined. Remember that, in our body-mind system, feeling is what joins body and mind. Well, parallel to that, in the larger, cosmic, space, feeling is what joins energy and awareness.

And by the way, these are not just some weird theories that I'm making up. Although I'm trying to write in language that we all understand, what I'm saying is based on very traditional teachings of meditative traditions. The presence of awareness-energy-feeling filling all of space has been seen by skilled spiritual observers, through specific spiritual practices, in many traditions. (I write awareness-feeling-energy here because they really are one, looked at in three different ways.) You can discover these things for yourself in your personal experience. And it is to help you discover this that I'm writing these letters.

All of this is jumping ahead a bit, so don't worry if it seems difficult to follow right now. It certainly is very condensed and will take some explaining. But I thought I should say these things now so that as we go on you will have some sense of where we are heading.

In the next few letters I'm not going to suggest that science has *proved* that awareness-feeling-energy fills all of space. What I *am* going to show, however, is that for the past fifty years scientists have been squirming with the uncomfortable feeling that awareness-energy-feeling is lurking there *somewhere*. And in Letter 17 I'm also going to tell you of observations, made by respected scientists, that *support* the idea that space is where THAT is lurking. These observations, of psychokinesis, precognition, and distance viewing, were all carried out in accordance with strict scientific methods, with honesty and integrity. Yet, needless to say, they have been ignored, or worse, by conventional scientists. So, let's begin.

First let's look at space and what might be in it according to science. I want to remind you that we carry around with us all the time an image-feeling of space as completely blank, empty, lifeless, feelingless, unaware. In other words, space is everything we associate with *death*. The space and time of Mr. Death is like the empty stage on which we (and scientists) imagine the world happening.

Absolute Space as well as Absolute Time were *inventions* of Isaac Newton, remember. But these beliefs condition our whole way of being in our world, far more deeply than the perspective lines condition our way of seeing (look back to letter 1). I hope it is really clear to you now, Vanessa, how much our way of being in the world and of relating to each other is conditioned by these kinds of beliefs.

And we feel that this lifeless empty space is filled with things, made of inanimate matter, just like your third-grade teacher told you. But what are things and what are they made of?

What about that red barn outside my kitchen window? Is that a thing? It looks pretty solid, although it's quite old now. Some of the red-painted shingles are falling off. Someone cut a little hole in the side wall for the many cats to go in and out. It has changed a lot over the years, but, yes, I certainly think of it as a thing. It is a complex thing, though, made up of parts.

Now suppose we take one of those shingles, is that a thing? Yes, of course, you say, what are you trying to get at?

Well, that shingle is made up of parts, too, isn't it? Yes, you say, and I'm taking chemistry at school so now I know where you're headed. The smallest parts of things are atoms, right?

Well, yes, but let's look at the atoms. Are they solid lumps of stuff? No! You know that too? OK.

The atom has a minute nucleus that actually contains almost all the stuff of the whole atom, as well as all its positive electrical charge. And around the atom a bunch of electrons are buzzing like gnats. The electrons carry a negative charge, so that the atom ends up being neutral, electrically, that is.

The diameter of the nucleus is roughly one ten-thousandth the diameter of the whole atom. If you were to imagine pumping up the nucleus to be the size of a marble in the center of an empty space as big as the Houston Astrodome, then the electron would be like a gnat flying around the roof of the dome. In between the electron and the nucleus is space.

For a long time scientists thought that the *nucleus* was the lump of stuff that couldn't be broken up any more. Then they discovered that the nucleus is made up of a number of smaller particles called neutrons and protons. So they thought *these* were the smallest particles of matter that couldn't be broken up any more. But, lo and behold! They made even bigger machines to collide particles into each other at huge energies, and they found that protons and neutrons are made up of even smaller particles, the *quarks*. And the size of a quark is about one ten-thousandth the size of a medium-sized nucleus. So now the *nucleus* appears to be made up of mostly empty space with minute quarks flying around in it.

A nucleus is simply a buzz of flying quarks, just as a cloud of gnats is nothing but gnats flying within a certain perimeter. The boundary of the nucleus is defined just by how far a quark can go. To use our marble analogy again: If the quarks are the size of a marble, they will be flying around in a space anywhere from the size of a large barn to the size of the Astrodome. But this time there is no hard core in the middle at all.

So now the nucleus is thought to be mostly empty space within the mostly empty space of the atom. Is there any end to this? As physics writer Heinz Pagels wrote, "No physicist I know would be willing to bet much on it."

Many physicists think that quarks cannot be truly elementary (for reasons I won't bore you with) and are already talking about "prequarks," the building blocks of quarks. It is unlikely that these can ever be found, because a particle-smashing machine needed to find them would need to be so big that it would be unbuildable. However, if prequarks *were* found, there is little doubt that we would find the *quarks* to be mostly empty space.

So, however deep we look into matter, we find yet more empty space. But is space really empty? The answer, according to physicists, is *not at all!* Back in 1928 a brilliant young physicist, Paul Dirac, was trying to find mathematical equations describing the motion of an electron in an electromagnetic field. And he wanted these equations to combine the quantum theory that was just then being formulated (and that he played a big part in formulating) with Einstein's theory of relativity, which had also only recently been announced. The 1920s sure were exciting times for us physicists!

Dirac did find these equations, and they precisely explained some experimental details that had been puzzling physicists. So they seemed to be pretty good equations. But they were also very peculiar.

The best way Dirac could come up with to explain his equations' weirdness was to say that there is a vast ocean of "virtual" electrons filling all of space. By *virtual*, he did not mean that they were not real but that they exist in a way that makes it impossible for us to detect them directly. So, since we can't observe them in our world, they are "virtual." We do not normally find these in our experiments because they are not in our "real" world.

Dirac proposed that it might be possible to make one of these virtual electrons jump into our world from its virtual ocean. This would leave behind a hole in the ocean, and Dirac proposed that such a hole in the virtual world would appear to *us* as a positively charged electron. So he called it a positron. And, he said, if a positron did appear in any real experiments, then we should always find it together with the electron that had popped out of the virtual world (leaving behind the hole that looked to us like a positron).

Sure enough, Dirac's weird positron-electron pair was soon found by the experimental physicists.

The discovery of a particle predicted by theory is a very rare event. Such experimental confirmation of a theoretical prediction gives tremendous support for the underlying theory, although physicists were already pretty convinced of the correctness of Dirac's theory. So the discovery of the positron made Dirac's proposal of an ocean of virtual particles believable. *And this ocean of virtual particles is in what appears to us as empty space!*

But there's more strangeness yet! The electron-positron pairs are continually jumping out of the virtual ocean and disappearing back again.

And this appearing and disappearing act of zillions of electron-positron pairs leaves energy behind in our "real," or experiencable, space. This energy of so-called empty space, caused by the ocean of virtual particles, has been called the vacuum energy. It is also referred to as zero-point energy, because the energy would still be present at absolute zero temperature when there is no heat energy or other kind of energy at all.

So there is energy even in space that has no real matter in it at all. In fact, space is so full that according to some calculations a thimbleful of space contains enough zero-point energy to boil off all the world's oceans.

We saw that in this new view of matter the elementary particles themselves are empty of anything approaching our usual idea of substance. They seem to have no stuffing. But now we find that space itself is *no longer empty*.

For decades the vacuum energy was thought to be a very esoteric fact that applied only to calculations of very fine detail and was otherwise irrelevant to the real world. Recently, however, physicist Hal Puthoff and his coworkers have shown some remarkable things about this energy. They have shown that the basic states of ordinary matter are not inactive and static but are constantly interacting with the underlying, zero-point energy. In fact, the presence of this energy is *necessary* to sustain the structure of ordinary atoms. As Puthoff says, "Pull the plug on the zero-point energy and all atomic structure would collapse." Which means, of course, that our whole world would collapse.

In another study, Puthoff showed that the property of inertia that all bodies possess—that it takes a force to get them moving, and once moving, they keep on moving until another force stops them—is actually resistance to being pushed through the zero-point energy. Puthoff regards this as an extremely fundamental result in physics that provides a clear connection between the strange zero-point energy and the ordinary world of things.

Puthoff and his colleagues have demonstrated the existence of this energy experimentally and have even begun to design machines to extract the energy from the vacuum. So convinced are they of the workability of this project that they have even taken out patents on their designs.

Privately, Puthoff speculates that the zero-point energy is the physical manifestation of the "ubiquitous, all-pervasive sea of energy that undergirds and is manifest in all phenomena" that has existed throughout human cultural history and through which humans and the cosmos are interconnected. "This pre-scientific concept of cosmic energy," says Puthoff, "goes by many names in many traditions, such as chi, ki or qi

(Taoism), prana (yoga), mana (Kahuna), barakah (sufi), élan vital (Bergsonian metaphysics) and so forth."

A sixteenth-century Confucian sage, Wang Shihuai, would agree with Puthoff. He writes:

> The name "mind" is imposed on the essence of phenomena. The name "phenomena" is imposed on the functioning of mind. In reality there is just one single thing, without any distinctions of inside and outside and this and that. What fills the universe is both all mind and all phenomena.
>
> Students wrongly accept as mind the petty, compartmentalized mind that is vaguely located within them and wrongly accept as phenomena the multiplicity of things and events mixing together outside of their bodies. Therefore they pursue the outer or they concentrate on the inner and do not integrate the two. This will never be sufficient for entering the Path.

There are numerous stories of masters of Chinese *chi gong*, tai chi, Zen swordsmanship, and aikido demonstrating remarkable physical effects of spiritual energy. There is a film of Ueshiba, the founder of aikido, being attacked by several of his students simultaneously when he was in his eighties. As the students lunge toward him in a circle he suddenly appears, from one frame to the next (i.e., within one-sixteenth of a second) outside the circle and the students go crashing into one another. I have seen a demonstration by a Japanese master of Zen sword in which he simply held his hand, palm facing down, a few feet above the head of a volunteer lying on the floor. He asked the volunteer to move. The volunteer told me, "I was incapable of moving. It was not that I was hypnotized, but that I felt such a powerful force emanating from the man's hand and pressing me down."

*Ring of Fire* is a documentary video by two British brothers, Lawrence and Lorne Blair, who spent ten years traveling and filming in the volcanic island chain of Indonesia. In Jakarta, the capital of Java, Lorne developed an eye infection and went for treatment to an acupuncturist. The acupuncturist placed needles in the usual points, but then held them. Lorne reported that he felt powerful electric shocks through the needles and could not control the movements of his muscles. He was visibly twitching on camera. The doctor explained that he was sending *chi* through the needles, and after successfully treating Lorne, he was asked to demonstrate chi some more. First he touched Lawrence's hand, and after a brief pause, Lawrence pulled his hand away quickly, saying that he had felt a sudden shock of intense heat. Simmie, the sound engineer, looked especially skeptical, so the doctor repeated the demonstration for her. Simmie held her hand on his bare stomach and after a brief pause she pulled it away, with a look of astonishment on her face. For several seconds after, she was seen giggling and shaking her hand.

The doctor ended his demonstration by crumpling a newspaper supplied by the film crew and holding his hand a few inches from the paper, his arm rigid like a lightening conductor. After a few seconds' pause, a shudder went through the body of the doctor and down his arm. And the newspaper burst into flames. The astonished look on the faces of the Blair brothers and the film crew, quite apart from the humbleness and obvious ingenuousness of the doctor, leaves little doubt to anyone watching the film that this was not a stage production.

The doctor explained that anyone can generate this energy if he or she practices meditation every day. He also emphasized that it is important in working with this energy not to have negative emotions and aggression, because the energy is neutral and can be harmful to others rather than helpful if it is misused.

So how shall we imagine matter now, Vanessa? That "stuff the world is made of"! How shall we imagine our atom now—our little elementary impenetrable ball? Perhaps we could say that it is a surface, or boundary, determined by the electron orbits, in already full space. Then the nucleus is a deeper surface in full space, and the neutrons and protons . . . down to the quarks . . . down to what? all are surfaces in the fullness of space. By surface, I mean a boundary, something that divides one part of space from another. And these boundaries appear to us (whether or not they are there independently of us) through the interaction with the fullness of space of our sense organs, sight, touch, and so on—or their extensions with instruments. When we look with the naked eye we see the surfaces of trees, rocks, people, and so on. When we look with a microscope, a boundary is cut in space at a deeper level, and so on.

Matter is nothing but boundaries. Boundaries within boundaries within boundaries . . . in the _fullness_ of space. And what is this fullness? Well, we have seen that physicists believe that the fullness of space is energy. And in the next letter I'll show you that, try as they might, they can't seem to quite get rid of awareness either. You see where we are headed? Let's go on then—to awareness in space!

# LETTER 15

❦

# *Awareness, Space, and Energy Come Together at Last*

*Dear Vanessa,*

Yesterday we discovered that science seems to be telling us that the "stuff" the world is made of is space full of energy, and that the things we believe we see are surfaces in this energy sea. And we saw that this energy sea could well be another view of the energy of chi supporting the world that has been described by spiritual traditions.

In this letter I am going to try to explain to you the strangest thing of all about that beautiful new physics that was born in the 1920s and still underpins the physics of today. This strangest thing is that somehow the awareness that scientists thought they had shut out of nature several centuries ago has come back in. This is what blew my mind when I was a teenager reading James Jeans on a warm April morning. Yet, of course, it is only a return to a more whole way of knowing our world.

As we saw in the previous letter, elementary particles are nothing but surfaces in the energy ocean that is all of space. The creative interaction between awareness and that basic energy sea seems to go down all the way to how physicists observe elementary particles in their experiments. And experiments are the only way we know and the only way we can ever know about the elementary particles of physicists' stories. Let me explain why physicists seem forced to bring back awareness, though some don't like to *at all*.

First we need to talk about another strange, weird thing about this deep, tiny level of our world. This is that elementary *particles*, such as the electron, can sometimes appear to behave like *waves* (or fields, as physicists prefer to say). This altogether contradicts the picture we carry around with us of how the world is: full of *things*. A *thing* cannot be both a particle and a wave. It just can't, that's it. But electrons can! And so can all the other tiny particles.

Let's reflect a little on the basic difference between a particle and a wave. This will help us to understand why a thing in our human-sized world can't be both a wave and a particle, and why this behavior of electrons is so frustrating to physicists.

All the energy and mass of a particle is concentrated in a small region of space at a particular time. The TV is over there. Sernyi's toy is here. My chocolate cookie is there. Each is in a particular location; we could say they are *localized*.

The energy of a *wave*, on the other hand, is spread throughout space. If two people hold each end of a rope and one of them shakes it, for example, a wave forms all along the rope. Or if a gust of wind catches the edge of a field of wheat, sending a ripple across the field, then the entire field has a pattern of movement—waves of ripples. A wave is simply not localized in space, whereas a particle is.

How can you tell when something is behaving like a wave motion? Well, one of the fascinating things about waves is that they interfere with one another to form amazing patterns. For example, have you ever seen a photograph taken from the air of the waves on a lake, or around a harbor? They are not just straight waves; they form wonderful patterns. Or, to see this right now, squeeze your first and middle fingers together until there is a very tiny slit between them. Now put the slit up to one eye (close the other) and look through the slit at a bright light or at the sky. Do you see light and dark bands running parallel to your fingers? That is because light is wave motion.

Particles and waves are completely contradictory ways of spreading out matter or energy in space and time. We could ask about the *position* of a particle, but it would be meaningless to ask about the position of a wave. On the other hand, we can ask about the *frequency* of a wave (the number of times it vibrates per second), but it would be meaningless to ask about the frequency of a particle.

So, based on how things behave in our ordinary world, something simply *cannot* be both a particle and a wave. But electrons sometimes behave like particles and sometimes behave like waves! If you ask where an electron is, for instance, if you ask it to make a little dot on a photographic plate, it will do that. In other words it will say, "I am *right here!*" But if you send it through the electron equivalent of the slit between your fingers, you will see patterns just like those light and dark bands. The electron is telling you, "I'm spread out just like a wave. Ha, ha!"

If we try to think of electrons as ordinary particles like tiny billiard balls, having a definite position and a definite path through space and time, they keep telling us, "You can't think of us like that, see, 'cos we're waves as well." Whatever the electron is, it is nothing like a little ball of *stuff*. And the same goes for neutrons, protons, quarks, and all the other "elementary" particles that are supposedly the building blocks of the universe.

And what is it that changes the electron from being like a wave to being like a particle? It is *the fact that you observed it, or measured it.* This is a key point. Now what on earth (or in all the heavens) does all this mean?

Let's first ask Niels Bohr, who is often regarded as *the* father of quantum theory. He was also a wonderfully kind and brave man who refused to leave Denmark when it was occupied by the Nazis in the Second World War because he was able to use his physics institute as an escape route for Jews.

The idea that the world is made up of tiny particles that are localized in time and space must be complemented, Bohr said, by a picture of the world as made up of interacting fields (waves) spread out over all of space and time. The two pictures, according to Niels Bohr, are *complementary*. By this he meant that both are necessary for a complete description of matter, yet only one of the two pictures can be how things "are" at any particular moment.

Niels Bohr saw that physicists had to understand physics within a bigger perspective that included *meaning*. He intended his principle of complementarity to apply to life in a much broader arena than just quantum theory. He felt that the Taoist yin-yang symbol was a clear expression of the complementarity of all things in life. And when he was made a baron by the king of Denmark in honor of his work in physics, and his bravery in the Second World War, he incorporated that symbol as the central emblem of his new family crest.

Bohr felt that the principle of complementarity governed his own life, and this is beautifully illustrated by a conversation recounted by the psychologist Jerome Bruner. "Bohr told me," Bruner wrote, "that he had become aware of the psychological depths of the principle of complementarity when one of his children had done something inexcusable for which he found himself incapable of appropriate punishment. He said to me, 'You cannot know somebody *at the same time* in the light of love and in the light of justice.' " Love and justice, for Bohr, are *complementary*.

Bohr believed that the principle of complementarity also applied to the realm of consciousness. He wrote that "the contrast between the continuous onward flow of associative thinking and the preservation of the unity of the personality [is analogous] with the relation between the wave description of the motions of material particles, and their indestructible individuality." So, the nonstop flow of your thought is complementary to your sense of self. As long as you are one with the flow of thought, you cannot hold on to a fixed sense of your self. And if you try to hold on to that fixed self-image, then thought becomes frozen too, into repetitive habitual patterns. That is the basis of the cocoon.

Back to the electrons: after fifty years there is still no agreement on how to interpret the wave-particle duality at the depths of space. A variety of interpretations have been suggested, but none of them are really satisfactory. One interpretation says, "We don't know what is going on, so let's just ignore the problem." This is the attitude that most working physicists, and certainly other scientists, take. It is a cop-out, though, to an inquiring mind.

Another well-known interpretation, proposed by Nobel prize physicist Eugene Wigner, says, "It is the observer's *consciousness* that makes the electron switch from being a wave to being a particle." But Wigner and the others who support this view have little idea of what they mean by consciousness, or awareness. They concluded (just like Descartes) that mind is an altogether separate realm from the material world. So it just comes in to help solve the wave-particle problem like a Mr. Fix-it from outer space. Mind and matter are still completely *separate* realms and there is no real explanation of why and how they join. This, of course, is unsatisfactory for the physicists who want to keep mind out of the world altogether. It is also unsatisfying if you seek an understanding, a story, that heals the split between mind and body.

The other story that is the main competition to Wigner's is called the many worlds interpretation, and was invented by Hugh Bryce and Everett de Witt. The many worlds story is supported by another large group of physicists because, at first sight, it *seems* to keep mind out. This story says that every world that is *possible* exists right now. So you are actually happening right now in a zillion possible worlds. So am I, so is everybody and everything. In each world our lives are different—slightly or hugely—but everything that could possibly happen is happening, right now. This is certainly science fiction stuff, but it leaves out a crucial point—why you, and each of us, only experience *this* one world, here and now. And the answer would be because our awareness is experiencing it! So to make one real world, out of the zillion possible worlds, we still have to bring in awareness.

When confusions like this occur, they are confusions in human thinking. They are contradictions in our *thinking*, in our mental picture of the world, not in the world itself. And the way to resolve the situation is to go to a broader level of understanding that can accommodate both sides of the contradiction in one unified picture. It is the same in human behavior. When we feel conflict about something—liking someone at school and not liking him at the same time, for example—we have to go to a broader or deeper feeling for the situation, to see the person from a broader viewpoint. Then we realize that we can *both* like him *and* not like him, and appreciate him *beyond* liking *or* not liking.

This is how physicist David Bohm, a colleague of Albert Einstein, approached the matter (and the mind—sorry!). Bohm was well aware that, as I discussed in letter 10, thought, or language, and the reality that this thought is about do not really correspond, as we usually like to imagine they do. If we regard our theories as direct descriptions of reality as it is, then we will believe that there really are separate things corresponding to the various terms and distinctions we make in our theory. "We will thus be led to the illusion that the world is actually constituted of separate fragments," Bohm writes.

Relativity and quantum theory both suggest that physicists need to look on the world as an undivided whole. And in this wholeness all parts of the universe, including the observer and his or her instruments, merge and unite. This wholeness, the totality of our world, is dynamic, flowing. And "things" can be seen to form out of this flowing wholeness and dissolve back into it.

Through our thinking and language, humans make distinctions within wholeness that divide things one from another. Mind and matter, said Bohm, are both products of thought. Long, long ago we separated mind from matter by our thought and language, and then we began to believe they are separate in reality. Having invented them as separate stuffs, our thoughts then try to decide whether they are the same or different. If we confuse our thoughts with wholeness itself, then we and our world become fragmented and confused.

Awareness, as well as matter, is enfolded within wholeness. And our normal sense of space and time, as well, must be thought of as enfolded within wholeness, for these too are products of thought, inventions, not realities. What does Bohm mean by *enfolded*? Well, he gives an example of a jar of some sticky fluid, like glucose. Now, if you drip a drop of ink into the glucose, it will just sit there as a drop, it won't mix up. If you carefully stir the glucose round and round, say twenty times, the drop will gradually spread out until it seems to disappear into the glucose. We could say the ink is *enfolded into* the glucose. But the amazing thing is that if you now stir the glucose twenty times in the opposite direction, the drop of ink will reappear. The drop is *unfolded out* of the glucose.

Bohm uses the analogy of the hologram to suggest the way all the things we perceive, from elementary particles to galaxies to thoughts and emotions, are enfolded into wholeness. A hologram is a pattern on a photographic plate that is produced when two laser beams converge on it after one of the beams is reflected off an object. The plate can be read back when a similar laser beam is shone on it, producing a three-dimensional image of that object.

The amazing thing about a hologram is that each small area of the plate contains the information to reproduce an image of the entire object (the smaller the area, the fuzzier the image, but a complete image can be generated from even a very small section of the hologram). Whereas in an ordinary photograph, of course, each small area of the print contains just a small part of the image.

Wholeness enfolds all things in our world in some similar way to the way the holographic plate enfolds the three-dimensional image. Just as all of the image can be recovered from any small region of the holographic plate, so all of reality can be revealed within each tiny part of wholeness. Saint Teresa is said to have held an acorn in the palm of her hand and exclaimed that she saw all of reality in that small acorn. We have seen this principle, as well, in the idea of medieval alchemists that the microcosm, that is the human body-mind, is a reflection of the macrocosm, the whole world. Or, as they said, "as above, so below."

A moment of experience occurs when a particular patterning is unfolded out of wholeness by our perception. At ordinary human levels of patterning, when you hear a loon call, your hearing draws out the sound of the loon's cry from within wholeness. At microscopic levels, the sight of a bacterium on a slide is drawn out of wholeness by our act of looking. At subatomic levels of patterning, whether we observe an electron in the form of a wave or of a particle depends on how we draw that electron out of wholeness through our experimental instruments. And at yet other levels, the gods are drawn out of wholeness by our calling on them through ceremony and ritual.

You should not think of the undivided flow of wholeness as some otherworldly dimension, or something abstract or "mystical" that we can't really know. That is not the point at all. Wholeness is your real, immediate experience, *right now*. It is that experience as a whole, before your thought and perception break it up into little pieces. You cannot *consciously* perceive wholeness, because then you would be dividing it into "you" and what "you" are experiencing, so it would no longer be wholeness. But you can know wholeness directly, intuitively, when you stop dividing it.

You can know wholeness, says Bohm, when thought comes to an end. That does not mean that you stop thought but that you are no longer trapped in your thought. You are no longer trapped into believing that the fragmented reality that thought *wants* you to believe in *is* reality, wholeness.

For many years Bohm was a close friend and associate of the spiritual teacher Krishnamurti. In his later years, Bohm applied the understanding that he had gained through physics and through his dialogues with Krishnamurti to help students come to the discovery of a more profound dimension of their lives.

His last book is a series of dialogues with students that took place two years before his death. In it, he shows how thought enters deeply into our ordinary perceptions. He suggests that everything in our usual manner of experiencing is thought, including perceptions, interpretations, emotions, and even body. The fundamental flaw in thought, Bohm suggests, is that thought does not know that it is profoundly involved in perception. Instead, thought tries to make us believe that our perceptions are really outside of us and that we are separate from them and watching them. Thought convinces us that our automatic unexamined perceptions tell us how things really are. This will all seem very familiar to you by now from my previous letters, Vanessa.

Bohm points to a broader dimension of thought than the merely individual—it is one undivided system that pervades all human culture. Because thought refuses to see its own process, it has to make us believe that it has a source. And that source is the ego, or cocoon. Thought leads us to believe in a self-image that is itself merely a figment of thought. And because this self-image is a construction, it has to be protected by thought. So the processes of thought are constantly distorted by the need to protect the self-image and the assumptions attached to this self-image.

In Bohm's terms, because of this basic contradiction between our perceptions and what is, our actions become incoherent and the world becomes more and more chaotic.

The problem of thought, according to Bohm, can never be overcome merely by further thought. There has to be some ending to the whole process of self-deception; an ending not to thought itself but to the tendency to believe that the projections of thought are real.

Bohm has an unusual and very illuminating way of describing this process of ending the flaw in thought. He points to the sense of proprioception (literally: self-perception) which refers to the inner sensation of our body, the feeling of our body. With our eyes closed we know precisely where our arm is by an inner sense, a kind of livingness, warmth, or feeling tone. Tragic results follow when this inner sensation, proprioception, is lost. Yet the proprioceptive sense is so close to us that we usually ignore it.

According to Bohm, thought is a movement of the brain and, like the movement of an arm, we can detect it through a proprioceptive sense. We can become aware of the movement of thought at the moment it happens. If we do this, says Bohm, thought begins to see it's own flaw and the flaw begins to dissolve of itself.

All of this is also very similar to Buddhist views of thought and perception. For the Buddhists, the ending of self-deception is brought about by the practice of mindfulness-awareness, in which we become one with thought by being mindful of thought, perception, emotion, and bodily sensation as they are happening. It is thought itself that invents a separation between the thinker and the thought, so when this separation is brought to an end, perception is direct. We see what is, without the overlay of thought's interpretation.

Well, Vanessa, this has been quite a trip so far. So let me summarize where we are at the moment, in this question of the stuff the world is made of. We took your third-grade teacher's statement that this stuff is matter, and we said, "All right then, what is matter?" Seemed like a fair question to me, along with several generations of physicists.

So we asked the physicists what their "matter" is. And in the previous letter we ended up concluding that what we call matter is simply surfaces in space. The surfaces are determined by how you look: there are surfaces at the human level, like my red barn, that I see with my unaided eyes; surfaces at the biological level, like the cells of your skin, that you can see with a microscope; surfaces at the chemical level of the atom, seen in the electron microscope; surfaces at the level of the nucleus and elementary particles that can only be seen indirectly with atom-smashing machines.

Far from being the dead emptiness of Newton, space is actually the fullness that supports and energizes all these surfaces. Matter is nothing but surfaces in the fullness of space. Years ago I used to hear Chögyam Trungpa say, "Space is solid, the blade of grass is hollow." And I thought he was just talking metaphorically, though *he* seemed to be talking about actual perception. He seemed, in fact, to be talking about the reality of how things are beyond language. Well, in the context of the energetic fullness of space, I began to understand that Trungpa meant what he said quite literally.

Then we saw something even more weird about those surfaces that physicists call elementary particles. They sometimes behaved like particles and sometimes like waves, and in our human world one thing just can't be both. And we saw that any way physicists try to understand this, they seem to need to bring back the mind that their predecessors threw out so long ago.

Finally we realized, with the help of David Bohm, that the problem began with the way our thought, language, and theories divide up the whole of reality into little pieces (mind and matter, particles and waves) and then try to put them together again. Unfortunately, it seems that all the king's horses and all the king's men (as well as all the queen's physicists) couldn't pull it off.

And we realized that the wholeness that is beyond our thought and language, and beyond physicists' theories, includes awareness and energy, and surfaces of energy, as well as space and time. For all these are fragments created by our thought.

I have relied on David Bohm's way of describing wholeness in this letter because he was particularly articulate and went quite deeply into it. He was, as well, a first-class physicist and a colleague and friend of Albert Einstein's. But, Vanessa, don't get the idea that Bohm's was just some cranky view not shared by anyone else. Not at all. Many of the great physicists who thought deeply about these things in the years when quantum physics was being unraveled (the 1920s and 1930s) came to very similar conclusions. For example, in 1955 Nobel prize physicist Wolfgang Pauli wrote:

> To us . . . the only acceptable point of view appears to be the one that recognizes both sides of reality—the quantitative and the qualitative, the physical and the psychical—as compatible with each other, and can

embrace them simultaneously. . . . It would be most satisfactory of all if *physis* and *psyche* (i.e. matter and mind) could be seen as complementary aspects of the same reality.

And in 1988 biologist George Wald, who won the Nobel prize for his work on vision, wrote:

A few years ago it occurred to me that . . . seemingly very disparate problems might be brought together . . . with the hypothesis that mind, rather than being a very late development in the evolution of living things, restricted to organisms with the most complex nervous systems—all of which I had believed to be true—that mind instead has been there always, and that this universe is life-breeding because the pervasive presence of mind had guided it to be so.

After quoting physicists Arther Eddington, Erwin Schroedinger, and Wolfgang Pauli, Wald concludes, "What this comes down to is that one has no more reason to deny an aspect of mind to all matter than to deny the properties of waves to all elementary particles. . . . Mind and matter are the complementary aspects of [one] reality."

Although we have come to this point through the seeming abstractions of physics, wholeness is not some far-off, mystical reality, Vanessa, but is right here, right now. It is more intimate, more close even than space and time and the ordinary things we perceive in space and time. Because all of these, too, are fragments of reality split off from us by thought. And we can intuitively feel wholeness when we see through the fabrications of thought and the way it has distorted our perception and our feeling for so many years.

Now, we have come upon wholeness by asking about matter: What really are those little lumps of stuff the world is supposed to be made of? But we could have gone in the other direction and asked about mind: What are those little things supposed to be, in the head, that we call individual minds, or psyches? We will look at this in tomorrow's letter. And we will find that by going deeply into the psyche, we come again upon the realm in which mind and matter are found to have always been two aspects of one reality.

# LETTER 16

Mind and Matter Join in
the World Soul

*Dear Vanessa,*

We concluded at the end of my last letter that mind and matter are complementary aspects of one reality. We would expect, then, that we could come across this deeper level from the direction of mind, or psyche, as well as from the direction of matter. And I want to write a little about this today. It will help you to get more of a feel for the very personal, experiential quality of wholeness.

I am going to tell you about the work of Carl Jung and some of his more recent colleagues, particularly James Hillman, Marie-Louise von Franz, and Robert Sardello. I am going to try to do this in one letter, so this will necessarily be a very quick trip.

So, let's start with Jung. Carl Jung was a younger colleague of Freud. Freud was the first person, in the modern West at least, to recognize the presence of mental processes that affect our behavior but of which we are not aware—the *unconscious*. I emphasize "in the modern West" because very often in psychology texts you will find Freud being proclaimed as *the* discoverer of the unconscious. This is about as narrow and blind as saying Columbus "discovered" America. America was known to the native peoples for thousands of years before Columbus. Likewise, the presence of mind beyond conscious thought has been known and understood in unbroken traditions for thousands of years in Eastern cultures, particularly Buddhist and Taoist.

Freud did rediscover the unconscious in the context of modern materialism. After all aspects of mind had been withdrawn from the world outside and lodged in individual minds, in individual heads, it was inevitable that a deeper level of mind would have to be rediscovered. And, of course, since all mind was now supposed to be in the individual, that is where the deeper level of mind—the world mind, or world soul—would have to be discovered. Lo and behold, Freud discovered it!

For Freud the unconscious was entirely the repository of repressed instincts—particularly the sexual instinct and the instinct toward death. So from the viewpoint of the conscious ego, the unconscious was nothing but a threat. Jung broke with Freud on just this point. Jung realized that the unconscious went far deeper than Freud had been willing to admit. The unconscious, Jung discovered, held not just repressed instincts but a tremendous urge toward psychic growth and ultimate wholeness.

Jung discovered that, beyond the personal, individual level, each of us has a connection with a level of psyche that is common to all humans across all cultures. He called this level the collective unconscious. In his earlier years, Jung seems to have thought that this collective unconscious was still somehow located in the individual psyche—its collective aspect was inherited but still located in each individual. However, he later came to realize that the collective unconscious really could not be localized and individualized in that way, and that it is the *same* collective unconscious that each of us taps in to.

We should not think of the collective unconscious, or the psyche or the ego, for that matter, as *things*. Think of them more like dynamic energy processes. (I will be writing later about how primitive "thing thinking" is. But for now remember the Mi'kmaw people, whose language had only verbs and no nouns, as I wrote about in letter 10.) These processes, or energy patterns, are embodied in the individual body in the case of the ego, and in the world's body in the case of the collective unconscious.

The collective unconscious is a very hard concept for people clinging to narrow categories of thought, particularly the subject-object way of looking at the world. Is the collective unconscious subjective or objective? Is it only "in your mind" or is it "out there?" For the Western mind, these are the only two alternatives.

Yet the collective unconscious is not *objective*, because each person feels it as continuous with the conscious ego psyche. After all, it can be discovered, by each person who takes the journey that Jung did, through going deeply into the individual psyche. At the same time, the collective

unconscious is not purely *subjective*, precisely because it is colle
common to everyone, and whereas the personal ego is associated
embodied in, the physical body of each individual, the collective un
scious is embodied in the whole world.

In his last and greatest work, Jung muses, "It may well be a prejudic
to restrict the psyche to being 'inside the body.' In so far as the psyche
has a non-spatial aspect, there may be a psyche 'outside-the-body,' a
region so utterly different from 'my' psychic sphere that one has to get
out of oneself . . . to get there."

For Jung this region of the psyche outside the individual body was like
the world that traditional peoples, such as the Australian Aborigines,
experience—a whole world, invisible but present and interwoven within
this one, and inhabited both by the spirits of the ancestors and by the
gods. This realm could also be felt as a psychic realm, an "inside." In
other words, toward the end of his life Jung had begun to think, not in
terms of two worlds, but more in terms of two aspects of one world, one
inside and the other outside, a microcosm and a macrocosm.

The collective unconscious shows itself to the conscious ego, not directly, but in *images* that occur in dreams and visions. Often such images
have a timeless quality—they show up again and again from person to
person. Jung found that very similar images occur in the myths and stories of the gods in all cultures, and this was how he discovered the collective unconscious. He called these images archetypal. (*Archetype* comes
from a Greek word that literally means "first-molded" and means the
first, or highest, of its kind.) The archetype itself is a particular pattern
of energy in the psyche that cannot be experienced directly but only
through the archetypal images.

This psychic realm that is neither wholly inside nor wholly outside
is the realm where the gods, angels, dralas, daimones of all cultures dwell.
In fact, for Jung, the gods, angels, dralas, and daimones are images of the
more formless and primordial archetypal energies. And from that realm,
they have the power to affect the outer, physical realm—the human
dwelling place. The mythic stories of the gods are stories of this psychic
realm and show just how these forces reveal themselves on the material
level.

Just as we cannot say that the collective unconscious is merely subjective or purely objective, likewise the gods refuse to be pinned down in
that small-minded way. The archetypal gods do not exist in the way we
are used to things existing, as visible objects in the material realm. Yet at
the same time they do exist, independent of our individual minds, in that

ffect our lives profoundly—for better or for
e relate to *them*. By analogy, the wind is itself
r lives profoundly, cooling us on a hot sum-
omes in a hurricane.

es Jung found, which I will explain briefly,
..nd the *Self*. But there are many others: the
., ne Trickster, the Divine King, and so on.

...adow is that side of our psyche that is hidden from us and that
we tend to deny and suppress. But it nevertheless shows up in our world,
and the more we suppress it, the more it tries to make itself felt indirectly.
For example, someone might think of himself as a calm and reasonable
man, yet be prone to fits of uncontrollable and irrational anger that hurts
those around him. Someone else may think of himself as a passionate
lover, yet leave a trail of women heartbroken because of his cold and
unfeeling attitude to them. The calm and reasonable man may see everyone else as angry and irrational, and the passionate lover may perceive
everyone else to be cold and unfeeling. Unrecognized as part of himself,
the shadow has been projected away from him and onto his outer world.

The *anima* is the female aspect of a man's psyche (for women this
archetype is called the *animus*, and is the male side of her psyche). When
the anima is not recognized, it too tends to be projected onto objects in
the person's outer world, in this case usually onto women. So a man may
find himself constantly falling in love with a particular type of woman.
Each time, he finds himself unable to live without the latest version of
his ideal woman, desperately dependent on her, and miserable when she
is not around. This particular type of woman becomes a projection of the
man's anima and he will not be free from these painful dependencies
until he is able to find and unite with the anima within his psychic world.

The *Self* is the highest goal of each person. It is the full individuality
of a person who has grown into complete unity and wholeness. In some
ways the Self is an unattainable ideal, yet it is something for each of us
to strive toward. Images of the Self occurring in dreams often tend to be
various forms of a circle, the archetypal image of completeness and perfect harmony, like the mandalas of Tibetan Buddhism and Hinduism.
Jung saw how the Self drives each of us toward completeness, a process
that he called *individuation*. In the process of individuation each of us
becomes fully, genuinely who we are—a truly human individual. This is
very different from being *individualistic*, because being fully human
means realizing our inherent connectedness with our world—we become
whole not only in our own psyche; that wholeness includes our world.

Jung has become very popular recently, perhaps because the archetypal images, taken from myths and fairy tales, are so appealing. People like to see themselves as wizards, kings, tricksters, and so on. I, too, have described some of the archetypes very simply here, Vanessa, but I want you to realize that they are extremely powerful forces in the psyche that can at times overwhelm us. I don't think fooling around with little exercises to discover our "inner wizard" is quite what Jung intended. But by facing the archetypal forces in our own psyches, if we have the courage, and coming to know them deeply, we can grow to a new level of inner harmony and wholeness.

Jung made these discoveries about the nature of the psyche not by merely philosophizing at his typewriter but through intense experience with his own deep unconscious, as well as through fifty years of work with others. During one period of his life, lasting several years, he let go of conscious control and descended into the depths of his psyche so profoundly that he constantly feared for his sanity. Any less courageous person would certainly not have completed this dangerous journey but would probably have become stuck and then would merely have been considered mad by himself and others.

Jung pointed to a different sense of time that is experienced at moments when the archetypal energies, the gods, make themselves felt in our conscious life. Such moments often occur at boundaries, between sleep and awake, between conscious and unconscious, or even at the moment of crossing a physical boundary such as a bridge or crossroad. At these moments there is a sense of timelessness. Everything seems uncannily quiet, or seems to stand still. In the archetypal realm time is marked not by the passing of an endless series of empty moments—our usual sense of time—but by a continuous flow of energy felt as the ever changing quality of one moment, *now*.

For most of us in the modern world, such moments when we feel the presence of the gods are nothing but brief flashes. Our work in restoring the living world is to connect again and again with these brief moments and try to extend them. But for others, the world is seen simultaneously in both its mundane physical aspect and its timeless, archetypal aspect—recall the Australian Aborigines, or the Mi'kmaw, for whom a rock is just a rock but *at the same time* is an ancestral grandmother.

We often feel moments of timelessness or nowness when a meaningful coincidence occurs between two events—an inner, psychic image and a corresponding outer event. Jung called such coincidences moments of *synchronicity*. So let us look now at synchronicity, and this will lead us back to seeing that mind and matter meet at the level of the collective unconscious.

Startlingly often, Jung found that when patients were in a psychically highly charged state, they experienced an uncanny correspondence between an event in their psychic world (a dream or a vision) and an event in the outer, "objective" world. He observed this hundreds of times, both in his own life and in the lives of his patients, as did his colleagues.

The most often quoted case of synchronicity, recounted by Jung himself, concerns a young, well-educated woman who had a very one-sided logical mind and was stubbornly unresponsive to Jung's efforts to soften her excessive, harsh rationality. One day she was telling her dream of the night before: she was being given a golden brooch in the form of a scarab (a type of large beetle). As she spoke, there was a persistent tapping on the window—a large insect obviously wanted to come in. Jung opened the window and caught the insect. It was a golden-brown beetle, very much resembling the scarab from the dream. Jung writes:

> I handed the beetle to my patient with the words, "Here is your scarab." This experience punctured the desired hole in her rationalism and broke the ice of her intellectual resistance. The treatment could now be continued with satisfactory results.

Here are two more examples of synchronicity given by one of Jung's long-time collaborators, Marie-Louise von Franz, who made a special study of the topic:

> I go into a shop and buy a blue dress and have it sent to me. Three days later the package comes, I unpack it, and the dress is black. The dresses were mixed up by the shop. At this moment, I get a telegram saying that a close relative has died, for whose sake I will have to wear black. Naturally my reaction to this is that it is not just an accident, but that it is extremely "odd."

The connection here is through the meaning of the events. As Marie-Louise von Franz says, "In this example, the relationship in meaning is only through the blackness of the dress. We associate black with mourning, and that connects it with the external event."

This is very like the meaningful coincidences that I have already written about. Jung calls these events synchronicity rather than synchronous, or simultaneous, because they may not necessarily happen at precisely the same time. The objective event corresponding to the dream or image may not be verifiable until later. Von Franz illustrates this in the following tale about a patient who was extremely suicidal. Von Franz was concerned about going away for a vacation and did so only after extracting a solemn promise that her patient would write to her if she had any trouble. Von Franz relates:

> One morning, I was chopping wood and had these thoughts (I was able to reconstruct my thought process precisely later): This wood is still wet; I'll pile it up at the rear so that it won't be used first and will still have a chance to dry out. . . . Then suddenly I saw this patient before me and thought about her. This completely disrupted the flow of my thoughts. I felt directly how this other thing broke into my thoughts. I thought: What could that woman suddenly want from me? Why am I thinking about her. Then I asked myself if I could have somehow gotten the thought of her from my preoccupation with the wood. There was no associational path from the wood to her. I went back to the wood chopping and again her image was there—this time with a feeling of urgent danger. At that point I put aside my ax, closed my eyes, and thought: Should I take my car and drive to her immediately? Then I got a quite definite feeling: No, it would be too late. Then I sent a telegram: "Don't do anything foolish," with my signature. Later it came out that the telegram reached her two hours later. At that moment she had just gone into the kitchen and turned on the gas valve. At that point the doorbell rang, the postman delivered the telegram, and she was naturally so struck by this coincidence that she turned the gas valve back off and is now—thank God—still among us.

Coincidences such as these, *meaningful* coincidences, are clearly points at which the "outer," "objective" physical world meets the "inner," "subjective" world of the psyche, at the level of the collective unconscious. Jung is therefore forced to the conclusion that at some deep level, psyche and matter are one. He calls this level the *unus mundus* (one world). It is the level at which opposites such as mind and matter, subject and object, inner and outer, have not yet split. Jung writes, ". . . today we know this much beyond all doubt—that empirical appearance rests on a transcendental background [that is] as much physical as psychical and therefore neither, but rather a third thing, a neutral nature which can at most be grasped in hints since in essence it is transcendental."

Again I should emphasize, Vanessa, that the conscious ego, the collective unconscious, and the unus mundus are not separate, real entities that we can point to and say "Ah! There is the collective unconscious" or

"There is the ego." Rather, the conscious psyche, the collective unconscious, and the unus mundus are aspects of the whole psyche. They shade
into one another so that, at one end, the collective unconscious is continuous with the conscious individual psyche, and at the other end it becomes indistinguishable from the unus mundus.

*Unus mundus* is a term from medieval alchemy that meant the world
as it was in potential form, before it became manifest as the world that
appears to us. For the alchemists, unus mundus was the plan of the actual
world that was already present in God's mind before he created it; for
Jung, unus mundus was the primordial soup, or basic stuff, out of which
everything comes into existence. According to Marie-Louise von Franz,
Jung thought of the unus mundus as a single energy (though energy
doesn't seem quite the right word, being too much aligned with the physical aspect). This "energy" manifests itself at lower frequencies of vibration as matter, and in more intense frequencies as psyche. This, von
Franz says, "in many ways resembles the Chinese idea of chi." And she
also suggests that we could identify it with the implicate order of David
Bohm.

To understand how two events in mutual synchronicity arise out of
unus mundus, we must return again to the archetypes. Remember that
the archetypes *themselves* are not the images that appear in dreams or
visions, in fairy tales and myths. The archetypes themselves are the powerful forces, residing in the collective unconscious, or the unus mundus,
which manifest to our consciousness as images or visions. When the
archetypes have sufficient intensity, according to Jung, they can break
through into the realm of the conscious psyche and at the same time into
the realm of matter. And this is when events appear to have synchronicity. So the phenomenon of synchronicity provides a direct, experiential
demonstration pointing to a deep unity of mind and matter.

Jung saw that there was a kind of deep knowing to unus mundus
and the collective unconscious. This knowing, or luminosity, was a felt,
intuitive knowledge, which he associated with the greek goddess Sophia,
Wisdom. He also, at times, called it abstract knowledge because it does
not depend on ego consciousness or the outer sense perceptions. It is very
close, I think, to what I have been calling feeling-awareness. It is this
wisdom that feels the meaning in the archetype, or by which the archetype carries and knows its own meaning. So when an archetypal energy
enters our conscious psyche and affects our life powerfully, it does not do
so randomly or merely accidentally. It does so in accordance with that

deeper wisdom that is ultimately directing each of us toward individuation.

Abstract knowing, wisdom, is how we feel the meaning in meaningful coincidence. As indications of this kind of knowing, Jung pointed to experiences that we now call out-of-body experiences. In an out-of-body experience, someone in a deep coma from an accident or an anaesthetic is nevertheless able to perceive in detail what is going on around him or her, and this is often associated with the the sensation of being outside the body. Though ignored or scorned by scientists, these experiences have been reported so often by sane and reasonable people that it seems neither sane nor reasonable to ignore them.

There is another aspect to Sophia that I want to mention very briefly here. Sophia is the wisdom, the energy-feeling-awareness, of the one world as a whole, including the earth and all that lives on, in, and above it, and including also the sky with its sun, moon, and stars. Sophia is the wisdom of the world soul, the *anima mundi*.

This concept of the world soul is becoming increasingly prominent recently. Particularly James Hillman and Robert Sardello are pointing out strongly and clearly that it is imperative for us to pay attention to Sophia. The emphasis on individual therapy, caring only for our own souls, for the past hundred years has, they say, led inevitably to neglecting the world soul. James Hillman has even written a book with the title *We've Had a Hundred Years of Psychotherapy and the World's Getting Worse*, a title that says it all.

Robert Sardello takes up this theme in a deeply moving way. In a passage on grief, he suggests that the epidemic of deep depression that is sweeping our society should not be merely dismissed as disease. Rather, he says, it is inevitable that we must grieve; we grieve for the loss of our world. Sardello writes:

Grieving has become more open and recognized in many therapeutic endeavors over the past several years. . . . I believe that the prevalence of this emotion is something more than personal, something more than a psychological process belonging to individuals. Grieving is world-oriented, and those who have entered into this deep emotion are expressing what we all feel and are afraid to face. The activation of an active, conscious soul life in the face of the present world is necessary and unavoidable. We see grieving all around us, for this condition is us. Our Earth, this source of love, is dying; no measurements of science can tell us otherwise; no continual cover-ups by government, by industry, by the military, can hide what is felt in the immediacy of the soul experience of the body. As long as this is denied, what belongs to the world will continue to be interpreted as only personal psychological suffering, when it fact it is at the same time a world suffering.

As you know, Vanessa, I have for some time been quite critical of the new popular psychotherapies that classify sadness as a disease. As Sardello writes, it is far from being a disease but is a profoundly sane reaction to the degradation of our culture and world. This degradation is accelerating faster every day, and grieving for what we are losing is perhaps the first step to finding a new way of being in the world—putting the Dead World story behind us.

So now we have come full circle. In the previous letter we started out from matter and, by investigating it deeply, came to the conclusion that it is not ultimately separable from mind. In this letter we have found that if we start from the individual psyche—those little atomic minds supposed to be locked up inside individual bodies—and if we again investigate this deeply, we inescapably come, once again, to a realm in which the psychical world is not separable from the physical world.

Now, we might expect that if mind and matter, subject and object, awareness and energy, are one at a deep level beyond space and time, then there ought to be some way that this shows up in our experience. And this is exactly what happens. Meaningful coincidences and out-of-body experiences are already indications of this, as we have seen in this letter. And, as we will see in my next letter, there are now laboratory experiments, which are as well conducted and strictly controlled as any conventional scientific experiments, that also point to awareness-feeling-energy being spread throughout space and time.

# LETTER 17

## Suggestions of Mind in Space at Princeton

*Dear Vanessa,*

In the previous letters I have invited you to see that science has given us no reason *not* to accept our intuition that awareness permeates all of space. On the contrary, I have indicated that some very brilliant physicists suggest that this may be the best way to understand why nature responds the way it (or he or she) does when we ask questions about the fundamental stuff of the world.

Today I'm going to tell you about exciting and revolutionary work that is being done—*observations* that are being made—that do *strongly* suggest the intermingling of awareness and space. This work is being done by scientists who are highly respected for their work in conventional science, and it is being done with all the care and rigorous controls that they take with that work. In fact, it is done with far more care and controls than they would usually think necessary.

I am going to tell you mostly about work by people that I know personally, because that way I can personally assure you of their integrity and intelligence as genuine scientists. But there are certainly other people doing work of this nature with similar results.

In a small laboratory tucked away deep in the vast and prestigious Princeton University School of Engineering and Applied Science a quiet but dramatic revolution has been brewing for more than fifteen years. The laboratory is known affectionately to its operators and friends as

PEAR—the Princeton Engineering Anomalies Research laboratory. (Mom thought they should have capitalized the *L* in "laboratory," then they could call themselves PEARL.) It is presided over by Robert Jahn, professor of aeronautical engineering and dean emeritus of the School of Engineering and Applied Science, and managed by Brenda Dunne, who was trained in psychology and humanities.

Remember, Vanessa, that I visited PEAR for two weeks a few summers ago. I attended a two-week workshop they had organized to present their results to others. There were about forty-five participants altogether, including about ten faculty, scientists who in one way or another were doing research that related to consciousness.

The group was a remarkable mix of talent, sensitivity, and sharp insight. Almost all the participants had some science training, from transpersonal psychology to computer science and biochemistry. And many of them also practiced some kind of awareness training. It is rare indeed to find so many people together for two weeks with such a combination of openness and intellect. And it was a treat to be able to spend two weeks with them discussing everything under the sun, from bioluminescence (the fact that all living things emit light), in the formal sessions, to how to test out-of-body experiences, during the meal breaks.

The PEAR laboratory is investigating the role of awareness in establishing physical reality. They do this by studying the effect of the thoughts or intention of an operator on the behavior of random physical processes, like atoms decaying or balls falling down a pinball machine. Mind over matter, you might say.

The laboratory uses a very simple electronic machine called a random event generator, or REG. The REG electronically creates a completely random series of zeros and ones—it is the electronic equivalent of tossing a coin and counting the heads and tails. If you were to toss a coin a thousand times, you would expect to get very close to five hundred heads and the same number of tails. Likewise, in a series of a thousand zeros and ones generated by the REG, we would expect to get close to five hundred ones and five hundred zeros. And the probability that the series will differ by a certain amount from five hundred of each can be calculated exactly.

When the machine is just left running, without its being paid attention to by anyone, it prints out a random series of noughts and ones, just as expected by chance. The PEAR experimenters asked, What happens when someone (the PEAR group calls him or her the operator) tries to bias the output of the machine toward more ones or more zeros? The experimental arrangement is very simple: in a pleasantly decorated room an operator sits in front of the machine in a comfortable armchair and tries to influence the machine. The pleasant decor and comfortable armchair are important, Jahn and Dunne found, because it is helpful for the operator to feel at home in the laboratory and included in the experiment as a colleague, rather than being treated as a guinea pig, as so often happens in psychological experiments. They believe that this also helps to reduce the operator's self-consciousness, which sometimes seems to inhibit the processes being studied.

Jahn and Dunne deliberately did not choose operators who claimed to have special "psychic" powers of any kind. The were looking for indications of an interconnection between consciousness and the physical world, not for any special human talents. Over the course of fifteen years, hundreds of ordinary people have visited PEAR and taken part in these experiments, and thousands of series have been run.

The upshot of all this work is that Jahn and Dunne have established with a very high degree of certainty that *awareness can influence physical reality.* The intention of the operator can indeed affect the output of a machine. Taking into account all the trials over the fifteen years, the probability that the effects are merely chance is around one in a million. This would be a remarkably high degree of certainty for *any* scientific experiment.

When Professor Jahn first decided to set up these experiments in the engineering school, the administration of Princeton University was so concerned about bringing disrepute on the university that a special committee was set up that included prestigious faculty, senior administrators, and outside observers, many of whom were highly skeptical of the experiments. No funding was obtained from university or government sources and no publicity for PEAR was allowed. After many years of carefully watching the activities of PEAR and finding nothing at all to complain about, the committee was disbanded. PEAR is now accepted as a reputable, if still controversial, contribution to the life and work of the university.

The experiments at PEAR are very carefully controlled. There is no possibility for the operator to cheat or for some kind of hidden bias in the machine to create these results. In fact, Jahn and Dunne have considered every challenge by the skeptics and been at great pains to take all their concerns into account. They have expressed appreciation for the degree of skepticism that their experiments have met with, just because it has forced them to be extremely careful both in setting up the experiments and in being sure of their results before they publish them.

The results at PEAR do not merely show an overall effect of awareness, they also show fascinating details that serve to further confirm that these findings are not accidental.

First, the exact pattern of results is specific for each operator. That is, the way in which the machine deviates from chance shows a very characteristic pattern for each operator. And this pattern is seen even over the course of many repeated experiments.

Second, experiments were done with pairs of operators trying to influence the machine together, and the results were fascinating. The PEAR team found that the combined effect of two operators of opposite sex was significantly stronger than that of the individuals operating separately. And for a "bonded pair" (a married couple or a couple in a relationship), the additional effect was even greater—as much as seven times greater than that of the individual operators.

And third, experimenters also found a difference in response between men and women. The effect produced by men tended to be more consistent than that by women, but the women were able to produce a larger effect when they produced one.

All these results clearly indicate that the personal qualities of the operators are somehow involved in their ability to affect the machines.

These and more results of the PEAR work, as well as that of other laboratories, confirm that there is a very real effect here. The experiments have been reproduced by sixty-eight different investigators in a total of 597 experimental studies.

I want to mention one other experimenter in particular: Helmut Schmidt, who has been doing work similar to PEAR's for more than twenty years, beginning when he was an engineer at Boeing Aircraft. Schmidt was the pioneer in building the kind of machine used at PEAR, and it was his work that initially inspired Jahn to begin similar work at Princeton.

It has been estimated that when the results of all these studies are included, the odds against chance being the explanation for them are 1 in $10^{35}$ (that means 1 followed by 35 zeros; and to get an idea of the magnitude of this number, think that 1 billion is 1 followed by just 9 zeros).

There is no doubt whatsoever that some aspect of awareness, the intention of the operator, is capable of affecting the material world. Just *how* the intention of the observer is able to affect the machine in the

experiments at PEAR no one knows. But there is some indication that
the effect is happening at the level of feeling. The most successful operators very often report that the way they operate is to develop a feeling of
empathy or "resonance" with the machine. According to Jahn and
Dunne, typical descriptions of this feeling of resonance are:

> . . . a state of immersion in the process which leads to a loss of aware-
> ness of myself and the immediate surroundings, similar to the experi-
> ence of being absorbed in a game, book, theatrical performance, or
> some creative occupation.

> I don't feel any direct control over the device, more like a marginal
> influence when I'm in resonance with the machine. It's like being in a
> canoe; when it goes where I want, I flow with it. When it doesn't I try
> to break the flow and give it a chance to get back in resonance with
> me.

There we are, there's that old resonance again. Once you know it's
there, you find it everywhere. Not surprising; it is what makes the world
go round, after all.

These experiments are showing us two things. First they are showing
that mind is somehow able to influence matter. So mind and matter are
mixed in a very intimate way. We are not especially surprised, are we,
having got this far in our discussion? But it is exciting to see it in black
and white, so to speak.

But these experiments also show that mind can affect matter *over a
distance*. In fact, PEAR has tested this. They asked operators to try to
influence the machines from hundreds of miles away, and the effect was
just as strong. Feeling or intention spreading, or already present,
throughout space. Aha!

Now let's look at some other work that shows awareness crossing
space. The PEAR group investigated another phenomenon, known as
remote viewing. That is the ability to describe a scene from a distance.
This phenomenon was first investigated scientifically in the seventies by
Russell Targ and Hal Puthoff at SRI, a scientific research institute in
Stanford, California. I mentioned Puthoff, in letter 14, in connection
with his more recent and conventional physics on the zero-point energy
of space.

The remote viewing experiment goes like this: One operator remains
in the laboratory while another goes out to view a site. Neither the labo-

ratory operator nor the field operator (called the agent) has any advance knowledge of the site to be visited. This site is chosen from a number of sites that have been previously mapped out, assigned a number, and locked away. The number of the particular site for the day's experiment is chosen randomly by computer.

When the field agent reaches the site, which is usually about half an hour's drive from the laboratory, he or she simply scans the scenery with an open mind for a while. The agent then writes down a narrative description of the site and also answers thirty specific questions. Typical questions are, Is any significant part of the scene indoors? Is the scene predominantly dark? Is any significant part of the scene oppressively confined? Is there any explicit sound, such as auto horns, voices, bird calls, or surf noises?

Meanwhile, back at the laboratory, the operator opens his or her mind to the possible scene, makes notes of whatever he or she picks up, and also answers the same questions as the field agent. Later the two sets of answers are compared by a computer program.

The PEAR and SRI laboratories took extreme care to eliminate possibilities of fraud, influence by the experimenters, or any other factors that could have somehow conveyed information from the agent to the operator via known channels. The results of many experiments added together make it certain beyond any reasonable doubt that remote viewing is a legitimately observable and repeatable phenomenon that demands to be taken notice of by science. According to Jahn and Dunne, detailed statistical calculations show that in the PEAR experiments, the probability that operators acquired as much information as they did purely by chance is less than 2 in $10^{11}$ (a hundred thousand million).

An extra factor in these experiments is that the effect was just as great if the operator scanned the scene and answered the thirty precise questions *before* the agent rode to the site. These are the scientific experiments that demonstrate precognition that I mentioned in letter 5. Very similar results were found at SRI by Targ and Puthoff.

Hal Puthoff described to us the often quite humorous lengths to which his funders would go to make sure that he and Targ were not cheating. One time, for example, the investigator for the funder went out with the agent. They were supposed to stay at the scene for half an hour and then return to the laboratory. Fifteen minutes into the session, however, the investigator suddenly said, "Let's go," jumped into the car, and drove somewhere unplanned. Back in the laboratory the operator started describing the scene and fifteen minutes into it said, "That's strange, it suddenly seems to have stopped." Needless to say, the funders could not find any cheating!

By the way, the CIA has also conducted experiments in remote viewing. Needless to say, this was kept secret for a long time and was only recently made public. The work was examined by Jessica Utts of the University of California. She concluded, "That anomalous cognition [remote viewing] is possible has been demonstrated. This conclusion is not based on belief, but rather on commonly accepted scientific criteria."

I had an interesting lesson in the way our chattering mind can interfere with our intuition while taking part in one of these remote-viewing tests during the summer workshop.

In this test, Brenda Dunne went off somewhere unknown to us. We sat around in the lounge area of the conference facility and tried, in whatever way we could, to "see" where Brenda was. This is what I got: First, I felt the presence of greenery; it felt like grass or a number of small green plants. Then I had a sense of a huge plate glass window. And the place seemed very busy. I felt a lot of traffic going by me.

I immediately jumped to the conclusion that the greenery was grass or shrubs, the glass window was a bank, and I (Brenda) was standing on the sidewalk on a busy street outside a bank. The only problem that kept coming to me was that the traffic did not seem large enough to be cars. But since I could not figure out what else it could be, I left it at that.

Where Brenda actually had been was the produce department of a supermarket. Now, if you just compare the two statements "produce department of a supermarket" and "sidewalk of a busy main street outside a bank with grass in front," I would not score very high. But if I had just noted my fleeting impressions "greenery," "large plate glass window" "traffic that is too small to be cars" (i.e., the grocery carts), then I would have scored 100 percent.

By the way, just as a little note, as Brenda returned from her trip and was coming up the steps to our building, before anything had been said, one of the participants greeted her by shouting, "Produce department at a supermarket." When we asked him how he got it, he laughed, with a wave of his hands, saying, "Oh, my mother was good at this stuff."

The mistake I made is actually a very common one, so common in fact that Puthoff was able to use it to increase people's ability to score high, in other words to *train* them. The point was to catch the short flashes of intuition within the stream of interpretive chatter. And one of the reasons I have taken the time to tell this little personal story is to ask, Don't we do this all the time? Don't we constantly ignore our intuitive flashes about what is happening and go with our conventional interpretations? And this also tells us how we can train to be more in tune with our intuition—by learning to see through the constant conventional chatter that goes on in our heads. That is, by some practice like awareness practice.

Jahn and Dunne point out that if results are taken seriously by science, which they *must* now be, scientists must come up with a radically new view of the relation between consciousness and the physical world. (They say "consciousness" rather than "awareness," so I will stick with that here.) In this new view, consciousness would have to play a central role in defining material reality. This is a revolution in the making beyond relativity, quantum theory, or any other scientific revolution since Copernicus's time.

In taking first steps toward this new view, Jahn and Dunne turn to Bohr's complementarity principle, which I mentioned in letter 15. Niels Bohr said that the wave nature of the electron and its particle nature are complementary. By this he meant that both were necessary for a complete description of the electron, yet both could not be true at the same time.

Jahn and Dunne suggest that a similar complementarity must be involved for a complete picture of consciousness. We are so used to thinking of mind in a particle-like way—that each of us has *a* mind localized somewhere in our body—but to complete the picture, say Jahn and Dunne, mind must also have a wavelike aspect that stretches over time and space and interacts with all the other waves (minds) in that space and time. They work out a fascinating quantum theory of consciousness by strict analogy to the quantum theory of matter. You might like to ponder the analogy between this particle-field theory of consciousness and the characteristics of mind that I called mindfulness-awareness. There seems to be a connection, and I don't think it is trivial!

Jahn and Dunne are very cautious in building their quantum theory of mind. They are dedicated to staying within bounds that should be acceptable to the most strict of scientific critics. This is the tremendous value and importance of the PEAR work.

Whatever theory of mind and matter Jahn and Dunne come up with, they have no question that these results demand that we envisage some intimate relation between awareness (or consciousness) and the so-called physical world. It seems as if nature refuses to be squeezed into the narrow categories of thought that we humans constantly come up with. Is matter particle-like? Yes. But it is also wavelike. Is mind particle-like?

Yes. But it is also wavelike. Is space empty? Yes, but it is also full. And will this human naïveté, believing in our own projections, ever end? Yes, but it will also go on. And on and on.

The careful scientific experiments that the PEAR group and other researchers are doing strongly support the conclusion that we have come to in the previous letters, that awareness-energy-feeling fills the whole of space. We were led to that conclusion almost inevitably by asking physicists about matter, and hearing what they had to say about the tiniest pieces of matter that they make their world with. And we were led to that conclusion again when we looked deeply into the individual psyche with Carl Jung. There we came upon synchronicity and the world-soul, the unus mundus, more fundamental than psyche or matter. Now we have tangible evidence of it in the actual world of experience. In my remaining letters, I'm going to lift us back out of that all-surrounding, all-pervading awareness-feeling . . . back to the ordinary world of things, if you could still call that an ordinary world after all we have been through so far.

# LETTER 18

# *How Shall We Look at Things Now?*

*Dear Vanessa,*

We have gone as far as I can go, in these letters, in our discussion of the basic "stuff" the world is made of. Your third-grade teacher told you "matter is the stuff the world is made of." Mindless, lifeless. Well, I have tried to show you that the world is not that kind of "stuff" at all but is living, feeling awareness.

One thing I do want to emphasize here, Vanessa, is that the way to understand this is not just to try to think it, to try to figure it out, but to try to *feel* it. When you read over the previous letters again, try to read with head and heart together. Read with your feelings as well as your thinking mind.

But it would not do at all to leave you here, dwelling in the oneness of the energy-awareness ocean. I have not been writing these things to you just to encourage you to retreat from the world, to escape into an internal bliss. It is certainly true that spiritual practice can bring great bliss, because it helps us to realize and to feel our deep connection with the space of feeling-energy-awareness. Being deeply with ourselves, through intensive practice in solitary retreat, is important from time to time, to refresh and nourish ourselves and renew our connection with feeling-energy-awareness. But lifelong retreat is not what most of us need to do in our shared world today.

The world of red barns, black cats with only three legs, cars with flat tires, birth and death, is the world we are in, and that is what we need to work with. We need to connect with feeling-energy-awareness, or soul, as it shows *in* this world. In that way we can nourish awareness of the sacredness of our world. That is how to help ourselves and others. And that is how to begin to build a good human society.

So let's contemplate *things*. How shall we regard things, now that we know they, and we, are all *patterns* of feeling-awareness-energy, rather than *lumps* of lifeless stuff? Although every thing arises within feeling-energy-awareness and is a part of it, things nevertheless differ from one another. And the division into inanimate things, living things, and things that think does make some sense at the level of ordinary life. There are cars, rocks, trees, cats, humans, angels, dralas, and so on.

First, let's just look at the idea of a *thing*. We usually imagine our world as space filled with solid, real things. We think of the things as real and of the changes that happen to these things as secondary to their essential nature. Take Sernyi, for example. (And take her fleas with her! Just *kid*ding.) Over the years Sernyi gets sick, grows thinner or fatter, gets fleas, and so on. But we still think of her as essentially the same dog, the same Sernyi. We think of Sernyi as a real, solid, living thing.

This fundamental belief about our world comes partly from the way we take for granted that the world really corresponds to our language. The English language is full of nouns, so the world is full of "real" things, at least if we speak English. But how fundamental are things? Aren't they, too, a result of the creativity of perception? Couldn't we perceive our world differently, and so live in it differently?

Last night there was a howling storm. It did not seem like an angry storm, more playful. But the wind sure was loud! It had a rhythm—louder, softer, louder, softer. And it had tones, like an overtone singer I once heard who could sing two tunes, one very high and one very low, both at the same time.

Last night I particularly noticed how we call this pattern of weather *the* storm. I had been thinking about what to write to you about "things," how we make a thing out of every pattern in our life. Take the storm for example, is it a *thing*? No, of course it's not, you say. It's just a pattern of wind, snow, temperature, pressure, and so on. Remember how, just before I came to this retreat, I went through a period of being fascinated by the Weather Channel on TV? I would keep tuning in every few hours, watching the satellite pictures of an approaching storm, watching the maps with the moving lows and highs and fronts. The commentators would talk as if every part of the weather was a *thing*—the storms, the lows, the highs, the cold fronts and warm fronts, the jet stream.

I was particularly amused at the way they would make a *big deal* about some approaching storm, "a disturbance" they would call it. They would keep us informed of its progress for some hours or even days. Then the next day you would see the map with no sign of the storm or disturbance at all. And the commentator would never mention it again, wouldn't say where it had gone, or how it had gone. It had just gone. Those particular things, storms, sure do have a way of coming and going.

When we relax our obsession with things, to *feel* our world rather than try to possess everything in it, we can begin to feel the patterns of energy in it. We usually perceive certain of these patterns as relatively unchanging objects—trees, cows, rocks, mountains, galaxies—owing to the time scale of our perception. We project inherent existence also into rivers and forests, think of them as separate, independent things—although you would probably agree that rivers and forests do not really exist as independent entities. Williams Lake is not quite so definite a *thing* as our roses or the trees on our lawn or Sernyi.

A little further removed we project thingness into rainbows and clouds, but here we know even as we do this that there really are no rainbows and no clouds. They are just a result of how we perceive and name our world. Beyond this we give our galaxy a name, the Milky Way, knowing full well that it is really a group of billions of suns separated by vast distances.

Our awareness is woven into the things of our world through the creative role of perception. Remember all that we discussed, in letters 7 through 9, about this? That red barn is not just there, a fixed *thing*, waiting for me to see it like a camera does—we have already discussed that. Some surfaces appear to me as ordinary things of everyday life because of the particular nature and degree of sharpness of my perception organs. I see a red barn because the fullness of space interacts with my organ of sight in a red-barn kind of way.

If we change the resolution of our senses with instruments, say, for example, we look at something with a microscope, then we see an entirely new world. If I take that red shingle that has fallen off the barn and examine it under a microscope, I will not see a more-or-less flat red surface at all. I will see the individual particles of paint and wood. And if I look with an electron microscope, I will see the individual atoms. All these different surfaces in the fundamental energy ocean of space result from the particular way I am choosing to perceive them.

We think that the human is a thing, bounded by her skin. But suppose our eyes are sensitive to only a tiny part of the spectrum of light and other radiation. Suppose we had eyes that were sensitive to X rays. We would see each other as a skeleton with organs hanging from it. We would not see the skin at all, and if someone told us that he or she saw this layer like a bag that contained the skeleton and organs, we might think the person was nuts, because most people didn't see it.

And suppose there were another layer surrounding us, beyond our skin, that most of us cannot see? If anyone told us he or she saw this, we could think that person was nuts. Well, of course, there is such a layer. It is called an aura. Lots of people say they see it, and others *do* think they are nuts. But did you know that the author Michael Crichton, who wrote *Jurassic Park*, *Congo*, and all those other best-sellers, wrote an autobiographical book, *Travels*, in which he describes, quite simply and soberly, learning to see auras? He said he didn't much see the point of seeing auras, except that it made boring cocktail parties more interesting, as he looked around at everyone's auras! Still, he saw them. So is he, too, nuts? I don't think so.

We see thingness or inherent existence in some patterns rather than others only because of the scale of time and space intervals that our human body-mind is capable of attending to. If our perception changed so that we were able to detect changes only in time chunks of one year, rather than at the rate of about a tenth of a second, as we do now, we would impute inherent existence into entirely different things. Then a human would be a blip, lasting eight or nine "seconds" at most, barely noticed; a forest would be seen as a single changing thing, rather than as a collection of trees, snakes, and so on. A city, or even an entire society, might seem to be one thing, slowly changing. We could see continents bump together and move apart, or mountain ranges rise up and wear away. Perceiving at even larger time intervals, we might see a galaxy as one lump, rhythmically changing, like a living thing.

When people wonder, "Who am I, *really?*" they usually try to find a definition of themselves that does not depend on anything or anyone else. This is particularly true in our culture, the culture of the individualist. We don't like to see how we change, sometimes drastically, according to the company we are with: at school with the teachers you feel like one person; with your friends you might feel like a completely different person; and at home with Mom and me, someone else again. We feel that we are supposed to be a self that is unchanging and not dependent on our relations with others. Yet our self is defined so much by our relations. It can become very confusing, because we believe our self is a thing that we should be able to know as unchanging. Well, it's not a thing like that at all. It's not even an *it*.

So, awareness is interwoven with the energy sea, through our particular sense organs, to produce the things of your and my worlds. Our world comes into being out of feeling-energy-awareness when boundaries, surfaces, are created by the sense perceptions of our body-mind. You and I and the rest of the glorious human species share a world because we have very similar sense organs and perceptions. And we have language to label the world and smooth out the differences in our perceptions.

If we had different organs of perception, as animals do, another world would be there for us. Some animals, for example, rabbits, squirrels, and maybe even cats, can discriminate only two primary colors, in contrast to our three primary colors of red, blue, and yellow. Now suppose you could see only red and yellow, what would your world be like? It would be a flat-seeming world. There would be a range of hues from yellow to red with lots of different kinds of orange in between. It would not be quite like someone who was color blind, but it would not seem like the same world you live in now.

But at least a world with only two primary colors is imaginable. Now, other animals, such as turtles, pigeons, and ducks, see with *four* primary colors. We cannot even imagine what this would be like. (A teenage friend Alyssum Pohl once told me that when she was very young she used to think to herself that when she got older she would find a new color. She would imagine herself hoeing in her garden and digging up a big rock with a color that no one had seen before. She pictured this as her "job.")

Francisco Varela suggests that the way we might try to imagine how these animals see would be to suppose that the fourth color was like a pulsation, or flickering. So the other three primary colors flicker at different rates depending on the fourth color. It would be as if every surface in our world were pulsating at different rates depending on the fourth "color." So instead of seeing a flat yellow wall, as we do, one of these animals would see the wall flickering at different rates in places, depending on the fourth "color." But this really is just a way for us to try to picture it. It is not what it must be like for pigeons. We simply cannot *imagine* what it would be like.

What kinds of *things* does a snake see, with its infrared-sensing organ; what kinds of things does a bat see, with its ultrasound detector; or a horsefly, which has faster eyesight than ours and so sees our world in slow motion; or a snail that sees one image every four seconds and must see in a series of disjointed pictures (from *our* point of view, of course); or what things are in the ocean for a salmon? Their worlds are nothing like I can imagine. How can we say that these perceivers perceive the same *things* that we do?

One of the reasons that scientists are making such terrible mistakes now, as they start to talk about consciousness, is because, since the time of Descartes up to this very day, most scientists deny that animals have any feeling or personal "subjective" experience *at all*. It is hard to imagine that humans could have made themselves so numb, and dumb. What kind of beings are we turning ourselves into that cannot recognize the cry of an animal, elephant or mouse, laboratory rabbit or rat, as a cry of real pain, just like yours and mine?

So, back to things. What do we *feel* when we think of, or perceive, something as a thing? We unconsciously assume that a thing has its own separate existence, its own identity independent of everything else. A thing's essential nature, its "self," if you like, is separate and closed off from other things around it. It's not fundamentally dependent on other things for its existence. A rock is a rock; its rockness depends on nothing other than itself. If you go up in a spaceship and throw a rock into space, so that it has no connection to anything else, it will still be that rock, at least for quite a while. A tree is a tree, and although it needs water, sunshine, and nutrients from the earth to stay alive, we usually think that, just like the rock, its *being a tree* does not depend on these things. That rock is our ideal thing. And that is what most people in our society seem to think they are, or strive to be—indestructible, indivisible, and individual as a rock.

Believing we live in a world of real things is a very primitive form of human thought. As I have mentioned before, scientists invent some *thing* with inherent, separate existence every time they find a new phenomenon they want to explain: to explain heat they invented caloric, and so on. And now DNA is supposed to explain life; neurons are supposed to explain mind. And unseen blips of energy that make patterns on photographic plates are quarks that are supposed to explain everything.

Scientists call a thing that is completely separate from everything else and all other energy in the universe "closed." A truly and completely closed thing could not actually exist, of course—it would be in a universe all by itself, since nothing could communicate with it. But it is an extreme that we can imagine.

At the other end of a closed-open scale, something that is completely open to its environment would not really exist either. (The word *exist*, by the way, literally means "to stand out from the background.") For example, consider a baseball-sized bit of air three feet in front of your nose. How is this different from the baseball-sized bit of air right next to it? It wouldn't make much sense to give a noun name to that bit of air. That bit of air is so completely open to its surroundings that there is nothing at all to distinguish it, and so we can't really say it exists separate from the air around it.

So what about everything else? Well, everything else, everything in the actual universe, is partly closed and partly open. Obviously, you say! Thank you, I am glad you are following so far. It would be helpful to have a name to refer to that particular aspect of things—the fact that they are partly open and partly closed.

What's the point of inventing a new name for everything in the universe, you say? Well, when I was writing about language, I mentioned how giving a name to a feeling we have never named can make us more aware. I mentioned the Japanese word *yugen* as an example. Well, inventing a word to point to the fact that every *thing* in the universe is both closed and separate *as well as* open and connected to every other thing also gives us a lot of insight into our world. Hold on, you'll see.

In the past people often used to use the word *system*. They would talk about closed systems and open systems. And a lot of interesting ideas grew up around the study of the openness of systems. But *system* has become rather an unpleasant word. It is associated with authority, rigidity, and lifelessness—the school system, the political system, the prison system, a mechanical system, and so on.

There is another term, though, that has not yet been ruined by the thing-making attitudes of our primitive minds. This is the word *holon*, suggested by author Arthur Koestler in the 1960s.

The *hol-* part comes from the word *whole*, so this refers to the fact that everything in the world can be considered a whole, in its own way. Even the bit of air at the end of your nose could be considered a whole, though it would not be very useful to think of it like that. The wholeness aspect of a thing is its self-containedness, so in that respect it is closed. For example, take one of my rosebushes. It is definitely a self-contained thing in some ways. I planted that one rosebush in that particular spot. That rosebush blooms in yellow, whereas the other is red. That rosebush looks more healthy than this one. And so on. It is a whole.

But the ending *-on* makes a *holon* also a part (like electr*on*, neur*on*) of something else, part of a greater whole. And insofar as everything is part of a greater whole, actually many greater wholes, it is open. For example, the rosebush is connected to the earth and the rain and the sun. If it weren't, it would quickly die. It is also connected to me and you. We smell it, we prune it, we water it when the weather is too dry, and we put straw around it in the autumn. So the rosebush is also open.

You, Vanessa, are certainly a holon. You are whole, you are definitely you, Vanessa. But you are also a part; you are part of the Hayward family; you are a part of a particular circle of friends that hang out at Tony's coffeehouse; you are part of the Shambhala community; you are part of the human community; you are part of the community of animals and plants on this earth; and so on. All these communities are larger holons, by the way.

So every *holon* is a whole-part, and this is the same as saying it (you, he, she) is closed-open. Everything is partly closed because it has a surface, a boundary. And within this boundary it behaves like a whole. But because it is also partly open, connected with other holons around it, it is also a part of a larger whole. Actually we should say it is part of a larger *holon*, because that whole that it is a part of is, in turn, part of a larger whole. So the whole universe is seen, this way, as being wholes within wholes within wholes, or parts within parts within parts—in other words, holons within holons within holons.

Well, I think we'll leave it there for this morning. In these last few letters, I'm trying to ask how might we look at the world, knowing that everything arises like a fabulous dramatic pageant in the ocean of feeling-awareness-energy. And this morning we began by finding a new way to look at the things of our world, seeing them as holons. Thinking in terms of holons rather than things reminds us that everything is both open and closed, a holograph-like patterning of the feeling-awareness-energy ocean.

As we come back to the ordinary world of things, after having plunged into the glorious ocean of awareness-feeling-energy, I want us to see the surfaces of that ocean without losing the feeling of livingness. That is why I want you to think/feel that your world is full of holons, not things. When we say the word *thing*, we usually think of something static, unmoving. When we think of Sernyi as a thing, even though we may imagine her running across the lawn, we usually don't think of her as a buzz of inner motion. By contrast, I want to ask you, whenever you read the word *holon*, to think/feel a whir of motion. Think/feel a storm, or a buzzing swarm of gnats in the evening summer sunlight; think/feel a group of people dancing in a meadow. That's much closer to what a holon is—any holon, from your body-mind to a rock. So when you read "holon, " think/feel "vibrating, buzzing, quivering, radiating."

Oh, and one more thing, Vanessa, since we are talking about things: don't make the mistake of thinking that energy-feeling-awareness is a thing! It's not exactly a holon, either. It has no boundaries and is in the very large as much as in the very small. You might want to compare it with things that people have given names to, such as God, and so forth. Some people, such as the Taoists have left it unnamed. As soon as you give something a name, you start to think of it as a thing, separate from you. We have seen this tendency of the human mind many times in this letter. And then, if it seems very big and powerful, you start to worship it and make yourself feel very small and impoverished. That's the danger. So it's not a thing, and has no name, even though I have been calling it energy-feeling-awareness in these letters. I expect you can see by now that energy, feeling, and awareness are simply qualities of space. And space ain't no *thing*!

This afternoon I'm going to write some more about holons. I am going to write about what it *feels* like to be a holon. Then, in my last few letters, I'll write about the variety of patterns of holons in our world, some living, some thinking, all in a constant creative play of resonance and ritual.

# LETTER 19

❦

# *What Does It* Feel *Like To Be a Tree?*

*Dear Vanessa,*

This is really part two of the letter introducing you to holons that I began this morning. So let's continue with that for a while.

One of the very important things I want you to know about any holon is that we can look at it from two points of view. We can look at it from an outsider's point of view or from its own point of view. These are often called the "third-person" viewpoint and the "first-person" viewpoint. What is meant by this is that any holon is both an object of someone else's experience and the subject of it's own experience.

Let's take an obvious example of a holon—Vanessa. You can be seen from two viewpoints—mine (or anyone else's) and yours. From my viewpoint you are a young woman, out there in my world, whom I love and hope will live a full and happy life, who is delightful and smart and humorous and darn stubborn, who sometimes irritates the hell out of me (and vice versa), and so on. From your point of view, what can I say? I can't really say anything other than that *you* know what it feels like to be you. Only you know what it feels like to be aware of your Vanessa-ness.

You know what it *feels* like to be Vanessa. I know what it *feels* like to be Jeremy. And Sernyi knows what it *feels* like to be Sernyi. Often I wonder what it feels like to be Vanessa, or what it feels like to be Sernyi, but the only holon that I can *usually* know what it feels like to be is Jeremy.

And there is a big point here, that *every holon in the universe has a "what it feels like" aspect to it*. Not only does it make sense to ask, "What does it feel like to be Vanessa?" or "What does it feel like to be Sernyi?" but it also makes sense to ask, "What does it feel like to be my roses?" or "What does it feel like to be that silver birch tree just outside my office window?" or "What does it feel like to be that big rock sitting on the hill beside our house like a protector?" or "What does it feel like to be the incredible storm of wind and freezing rain blowing outside my window?" (I don't mean what does it feel like to be *in* that incredible storm, by the way. I know what that feels like, and I hope you are not out in it now back up in Halifax.)

In other words, Vanessa, every holon in the entire universe has a certain degree of awareness-feeling. Every holon in the universe is a kind of gathering place or collecting pot of awareness-feeling. Or you could say that every holon in the whole universe is like a magnifying glass, but instead of focusing the rays of the sun, it is focusing awareness-feeling. If you imagine awareness-feeling to be like a flowing river, then every holon in the entire universe is like a big or little whirlpool in that river.

The complexity of their different patterns give holons different qualities of awareness-feeling: inanimate, like a rock; living, like a tree; emoting, like a dog; thinking, like a scientist. We assume, though we may be wrong, that not every holon in the universe has the kind of intense awareness that humans do (at least some humans, when they are really being human), or that dogs do. Trees seem to have less intensity of awareness, or at least a different quality of awareness; rocks, even less.

Dralas, gods, and so on, may have quite intense feeling awareness. Shamans and teachers who have direct access to the dralas tell us that, in fact, our fear of their intensity is the main thing that keeps us from fully seeing them. Even the smallest experience of dralas, or gods, may be shocking and disturbing to our security of mind—they shake our sense of normal reality and who we are, which can be terrifying. So, tuning in to the dralas is not necessarily comfortable or comforting—but it could be awakening.

You might be wondering what scientists would say about this inner, first-person view of holons. Well, since most scientists until very recently have tried to make us believe that even animals are not aware, you can imagine what they would think about the suggestion that *all* holons have an inner feeling—"mystical nonsense," they would huff and puff.

Not all scientists think that way, however. In fact, a lot of what I've been saying is very close to the ideas of a very great philosopher, mathematician, and scientist, Alfred North Whitehead, who lived from 1861 to 1947.

In the early period of his life, Whitehead wrote a revolutionary work on mathematics with the better-known mathematician and philosopher, Bertrand Russell. Then, in the 1920s, Whitehead went on to contemplate the profound changes that people of his generation would need to make if they were going to understand the newly discovered world of relativity and quantum physics.

I won't try to explain his ideas in detail, because it would take us too far from our topic. But the basic point is very simple: according to Whitehead, all events, things, and processes in the entire universe (all holons, to use our new term) are built from many feeling-energies coming together to form patterns of feeling-energy. For Whitehead, just as I have been saying, energy is the "outer" perspective of a holon, while feeling is the "inner" perspective. All feelings have some degree of awareness, and every pattern of feeling-energy has *some* awareness of itself as a center of feeling, however primitive that awareness may be.

Whitehead did not create just a vague philosophy based on his profound understanding of the importance of feeling-awareness. He worked it out in such detail that he was able to create a theory of matter, spacetime, and relativity very similar to Einstein's. In Whitehead's theory, of course, the world was built out of elementary *feelings* rather than elementary *particles*. Whitehead's work has been neglected by scientists partly because they don't *believe* in feeling-awareness. They don't train awareness to understand the messages of feeling. So people did not understand that Whitehead's ideas came from his direct experience rather than from mere abstract theory.

However, there is something going on in the scientific world today that would be truly revolutionary if it took hold. Some relatively mainstream scientists are beginning to say openly that awareness may actually be one of the basic elements of our world, along with space, time, matter, and energy. (Awareness is possibly even more basic than space and time, but few scientists have got that far.) It may have been a mistake to take mind out of nature so long ago. When Descartes took soul (feeling-energy-awareness) out of the material world, he may have set us on a sad sidetrack.

Here's how this potential revolution is shaping up. For the past twenty years or so (only that long!), neuroscientists have begun to ask, where in the brain is consciousness? The reason they haven't asked this before is that most of them were trying hard to deny that consciousness existed at all. Or even if consciousness did exist, they wanted to say that it had no importance whatsoever to the organism. Perhaps they thought that it was just a weird accident, I don't know. It is hard to understand what anyone is thinking who so denies the reality of his or her own experience.

Now, however, a big topic in the brain sciences is, what particular neurons, or group of neurons, are responsible for consciousness? As I have already suggested, it takes a pretty blinkered mind to be looking for consciousness in a part of the brain, but let's not go into that again.

Well, the people interested in consciousness now realize that even if they were to find a set of neurons that they decide are *the* neurons that cause consciousness, there is still going to be a major problem. There is a radical difference between (1) Francis Crick's knowing that a set of neurons in his own brain is making Crick himself conscious of seeing a rose, and (2) Crick's own *experience* of seeing a rose. Do you see what I am saying? Even if those neurons *are* doing the experiencing, there is a radical difference between saying *they* are doing the experiencing, and the actual experience of the "I" who is *doing* the experiencing. There is a radical difference between the firing of the neurons that make me see a rose and my *experience* of seeing a rose. You just simply cannot get from one to the other.

Another way of putting it is, even if I find out what every single neuron in a bat's brain is for and how it functions and interacts with all the other neurons, and all the rest of it, I *still* won't know what it is like to *be* a bat. Thomas Nagel, a well-known philosopher and neuroscientist, wrote a paper in 1974 called "What Is It Like to Be a Bat?" And this was one of the turning points that made scientists begin to wonder whether they had perhaps after all left something important out—"what it feels like." Of course, Nagel was only talking about beings that have conscious experience. As I pointed out in a previous letter, philosophers and scientists simply have a hard time telling the difference between consciousness and awareness because they are so hung up on thinking and frequently have little connection with feeling. Or they don't see what awake feeling has to do with "objective" reality, and think all feeling should simply be ignored.

This problem of how actual "I" experience, or subjective experience, could arise from neurons is being called the hard problem. And some neuroscientists are beginning to realize that it may not be possible to solve the hard problem at all, within the context of science as we know it. They are realizing, in other words, that subjective experience not only exists but is in a totally different category than an object in the world—just as, say, an apple is different from a joke. It was a mistake to take awareness out in the first place and we should put it back. So these scientists are wondering whether perhaps awareness should be thought of as a component of reality right at the beginning of everything, along with space, time, and matter.

This is certainly a huge step toward a return to basic sanity—by that I mean a return to a philosophy of life that begins and ends with actual human experience. It is not quite far enough, though, because you can't simply add in awareness as just another theoretical idea along with space, time, and matter. Awareness is altogether different from all of these just because it is first-person; our awareness is immediate experience, before any concepts or theories at all. To understand that awareness is *so* fundamental means a radical change in the way our society, and especially working scientists, even begin to think about what they are doing. As Francisco Varela writes, "This requires us to leave behind a certain image of how science is done, and to question a style of training in science which is part of the fabric of our cultural identity."

Do you see what a dramatic turnaround it would be if a large group of working scientists were to accept that awareness is irreducible (meaning completely basic)? The next question would have to be, "Well, how do we investigate this?" And the only possible answer would be *directly.* You can only investigate first-person experience, the "I" experience, with first-person observation. This means that scientists would have to do some direct first-person observation of their *own* minds, something like mindfulness/awareness. And they would have to allow observations of this type as valid observations within science.

This would completely change the face of science. Science could start to include phenomena that people have been experiencing for thousands of years regardless of whether or not scientists believe in them—phenomena such as those being explored at PEAR. And scientists might begin to include, in their training, practices that actually train them to be less influenced by personal bias and projections in their observations. Hey, wouldn't that be great! It would certainly make for better science. It might help to heal our world.

Now let's go back to holons, and the view from inside. Asking "what does it feel like to be . . . ?" gives you another way of understanding what I've been trying to say about awareness-feeling being throughout space. By awareness-feeling, I'm talking about "what it feels like to be . . ." Maybe we *can* feel what it is like to be another holon. We can reach out our own awareness-feeling to enter the awareness-feeling of that holon.

Of course, most of the time, most of us can only feel another holon very dimly. And that's why I have to wonder what it is like to be Vanessa, or to be Sernyi, or to be a rock. But sometimes we really do have a flash, a brief glimpse of what it feels like to be another holon, another person, or a tree, or a dog. These glimpses of insight are more frequent with a holon we really care for, that we have empathy for, that our feeling-awareness resonates with.

The holographic image for things, holons, can be quite helpful here. We could imagine, feel, visualize, that holons, ourselves included, somehow arise out of the space of feeling-awareness-energy like a holograph. So they are appearances in space. They are hollow because they are not separate from the space of feeling-awareness. Feeling-awareness space runs through things, within and without. When we see the hollowness of things, we can feel our direct connection with them. This is not just a cute idea, it can be actual experience.

Perhaps something like this has happened to you. A few years ago, while I was writing *Sacred World*, it was a sunny August afternoon and I had taken a pause in writing and was sitting out on our deck. Sernyi came running by, across the lawn. For a moment, just a moment, I saw her completely differently from how I usually see things. I cannot really put it into words, except to say that she was like an insubstantial light image moving through space; she was the same as space. And along with this brief glimpse of a different Sernyi came a quiet feeling of relaxation and joy, and a strong feeling of affection for Sernyi—yes, it's true, in spite of my constant bitching about her. It felt as if we were immersed in, part of, the same living, feeling-energy space.

These insights do usually occur only in brief glimpses. As soon as we start to think "Aha, I am having an interesting insight here," we immediately get self-conscious and lose it. But with some effort and practice, we can begin to recognize those moments more often and let them be there naturally. We have to be willing to develop our awake feeling organ, our awake heart.

Once, when I was about seventeen, a very big comet was due to appear one night. I went out with my father and we sat in the car on top of a hill and looked at the comet. And it was *so* frustrating, because you could only see it if you looked at it out of the side of your vision. It really was magnificent, and I really wanted to get a good look at it. But every time I looked straight at it, it disappeared. (Night sight uses the side of the retina, while daytime sight uses the center.) I kept having the uncomfortable feeling that I was not really seeing the comet.

Glimpses of feeling what it's like to be another holon are sort of like that—as soon as we try to catch them, they slip away. And this is also how it is with our experience of feeling-awareness-energy. We may catch a glimpse of it, in intensive meditative practice or in the middle of an ordinary activity. But as soon as we try to hold on to it, it goes. Our thinking veils it over for us once again. The key to stepping from the Dead World to the living world is just this—to practice again and again recognizing these short glimpses of feeling-awareness-energy and the nonseparateness of things, of holons.

Actually, it is sometimes helpful to begin, not by trying to feel another holon directly, but by feeling the space around holons. You could try this; for example:

Go outside into our garden, or if you are in the city as you read this, go to a park. Stand in one spot, pay attention to your breath for a few moments, and then pay attention to the sensation of your body. Now notice the space around your body. Let your feeling-awareness go out into the space around your body. Now extend that feeling-awareness to include the space in your environment. Stay there for a few minutes, letting your awareness go out into the space, and notice how you feel. Do you feel threatened, refreshed, agitated, nourished, healed? You don't have to put a name on how you feel—just notice how you feel. But if you do find it easier to give it a name, that's okay.

Now move to a second spot and repeat what you just did, putting your feeling-awareness out into space. Do the same in a third spot. Notice the difference in how you feel in each space as you move from place to place.

Now you can do the same thing in relation to particular holons. Go stand by one tree, put your feeling-awareness out. How do you feel? Now go stand by another tree, or a rock, put your feeling-awareness out. How do you feel now?

Here is something else you can do to begin to get the feeling of things. Choose an object you find particularly beautiful and care about. Put it somewhere free of clutter, against a plain background. Spend some time looking at the details of it. Let yourself appreciate its lines, its color, its texture, its quality. Now soften your gaze, relax your focus, and see the space around the object. You might notice an energetic quality to that space. Notice how you feel about the object and any change of perception when you relax your focus.

When you think of it, you can do this on the spot, with anything (any holon) your gaze falls on. First just look at the object in your usual focused way. Then, as before, soften your gaze, relax your focus, notice the details of the holon and the space around it, and notice any change in your perception of and feeling about the object. Notice, also, any completely off-the-wall thoughts or feelings that might flash in. Do you remember the incident with distance viewing that I told you about in a previous letter? And how it taught me to pay attention to the flashes and not to the story line? Well, it is the same here in trying to feel what it feels like to be another holon.

One important thing about all these little exercises that I am suggesting—*don't try too hard.* If you try, you'll think about it and wonder whether you are getting it, and you'll miss it, like trying to see a comet in the night sky. Just have fun.

# LETTER 20

## Holons Jumping Up and Down

*Dear Vanessa,*

In yesterday's letters I introduced you to the idea that everything in the universe is a holon, a whole-part. I emphasized the really key point that every holon has two perspectives—the "that" perspective and the "I" perspective. And I suggested how you might begin to investigate feeling the "I" perspective of other holons. What does it feel like to be a tree?

In this letter and the next one, I want to ask, what makes some holons *alive* in the conventional sense? And why do some holons seem to have thoughts and emotions? We'll see that the answers to these questions have to do with how the internal parts of a particular holon are organized and ordered. Now, when I say "ordered," I don't mean rigid and controlled, as in "law and order." A snowflake has a beautiful pattern to it and we say it has order. A living tree is more patterned than a pile of wood shavings. So we say the tree has more order than the pile of shavings. Order is the opposite of chaos.

Remember what Descartes said about animals, and even human bodies—that they are no different from machines. Well, until quite recently, scientists held to this belief like monkeys holding on to a banana. In fact they thought, or hoped, that everything in the world was like a machine. In letter 18 I wrote about closed systems, and a machine is just about the closest we can get to a closed system. So they developed a theory of closed systems, or machines. And this theory said that a completely closed system, completely isolated from its surroundings, would gradually and in-evitably lose its order. The arrangement of its parts would *inevitably* change from being more ordered to being less ordered, or more chaotic.

Let's take a simple example of how a closed system gradually loses its order. If you have a thermos of hot water and put an orange-juice pop-sicle shaped like a dog into it, the popsicle will melt into the water and you'll soon have a glass of tepid orange juice. On the other hand, if you put a popsicle stick into a glass of orange juice, you would never expect to see it separate into clear hot water with a dog-shaped orange popsicle in the middle of it. The glass with an orange dog in the middle of plain water is said to have more *order* than the one with the orange juice all mixed up with the water. The glass with water and popsicle is a relatively closed system, and if you just leave it alone it will become a glass of orange juice. It will go from more order to less order.

At the beginning of the century, it was believed that the universe was closed and eventually would lose all its order. This "heat death" of the universe was a topic of general fascination: all the energy in the universe would end up evenly spread through it, just like when we put an ice popsicle in hot water it quickly mixes to make tepid orange juice. Then nothing other than completely chaotic motion would continue for eter-nity.

When I taught high school physics in the early sixties, the cry "The universe is running down! The universe is running down!" was right there in the books we used, and it is still proclaimed in popular and elementary school science books with a kind of perverted glee. This story line has been incorporated into the fabric of our culture as one more factor in the general gloom about nature. Not only is nature fundamentally futile, nasty, and violent, but it is even sapping our energy and breaking down all that we create.

But all the systems in the world we *live* in are *fundamentally* open systems, they are holons. The earth itself is open, receiving constant infu-sion of energy from the sun and from cosmic radiation; the solar system is open; and the Milky Way, our galaxy, is open, as are all other galaxies and the galactic clusters and superclusters. Everything in nature, plants and animals, trees, flowers, birds, dogs, humans, organizations, whole societies, are all essentially open. They are all holons.

Scientists have only studied open systems, holons, seriously for the past few decades. This is partly because the mathematical tools have be-come available. But it is also largely because before this scientists really thought that the universe and everything in it were basically like machines, approximately closed systems, and so they didn't need to bother with open systems. Or so they hoped (because open systems are a lot more difficult to deal with mathematically than closed systems).

This recent work has shown that certain holons that already have quite a bit of internal order have an inherent capability to evolve to states of *greater and greater* order. That is, they organize themselves. The beautiful order we see all around us in the natural world is now allowed by the scientists' laws of nature!

A single cell of your body, for example, is capable of organizing itself—it feeds itself, builds all the parts of itself that it needs to live, and then divides every few minutes or hours. The holon that is you, Vanessa, organizes itself—your body-mind eats, sleeps, exercises, heals itself, learns to play the piano and do math, and so on. We say this kind of holon is *self-organizing*—it organizes itself.

A self-organizing holon exchanges energy and even order with its environment. Yes, holons can exchange order with their environment. For example, when you pick up your room, thinking about where to put things, you put order out from your body-mind into the room, and then the clean room puts order into your body-mind—you perk up. Or consider when you study the piano—through your practice, learning to read music, and so on, you are taking order into your body-mind. Then your body-mind is increased to a new level of order, so that you become able to play music at any time you wish.

A holon can maintain its wholeness and order, and even jump to *higher* levels of order. Here's how: a self-organizing holon maintains its level of order by constantly exchanging energy with its surroundings. For example, consider the whirlpool around a rock in a river. The form of the whirlpool (its order) is maintained by the constant flow of the river at a certain speed—water with a certain amount of energy is constantly flowing through the whirlpool. Now, suppose it rains heavily and the flow of the river increases, then the form of the whirlpool could radically change. So, when the energy coming into a holon jumps to a higher level, the holon can use this extra energy to create *more order* within itself. It rearranges itself internally so that it can *contain* the extra energy.

So holons can respond to shots of energy coming into them by raising themselves up, ordering themselves, to contain that energy. But there's another way a self-organizing holon might react to a shot of energy. When an extra shot of energy comes in, a holon can just collapse to a less ordered, more chaotic state. As an example of the two ways a holon can respond to an increase of incoming energy, think of your own body-mind, *feel* your own body-mind, as a holon rather than a thing. Feel the inner sensation of your whole body—now you can feel your body-mind's energy.

And you can feel different levels of energy and order in your body-mind. Sometimes you might feel very heavy, dense, lazy, fatigued. And often at those times you feel chaotic as well—your thoughts are running all over the place, your emotions are wild, and you certainly don't feel like picking up your room. At other times you might feel light, energetic, vibrant with life. And at those times you feel connected—body, thoughts, emotions, all seem to be synchronized, harmonized, *ordered*.

A shock to our body-mind holon, which is to say a sudden input of energy, can bump us up to a higher level of order, or knock us right back down to a low level. It could be a physical shock, such as breaking an arm; an emotional shock, such as ending a relationship; a mental shock, such as receiving a large inheritance. In each case, our body-mind can respond by raising up, energizing, embracing the shock; or we can respond by collapsing.

People who have been in a car accident, even a minor one, have noticed that the first moment of the accident they are tremendously awake—their perceptions are brighter, their experience of time slows down, their thoughts are clear, their emotions are strong. This is the first shock. From that point they can go one of two ways (provided they are not too hurt): they can feel enlivened and activated, or they can feel depressed, anxious, and confused about what to do.

Or suppose you are trying to concentrate on some schoolwork, or to read a book that you want to get absorbed into, and someone nearby suddenly turns on a stereo. How do you respond to that? Do you stop reading and just collapse into a state of irritation? Or do you use the energy to rouse yourself and focus even harder. Of course, if the music is too loud, it is very hard to keep going. But when the music is at a certain level, you do have the choice.

We can actually choose how we respond to shocks or sudden changes of our situation. Aren't we constantly faced with impermanence and transiency, so that sometimes we feel as if our life were one sudden change after another? If we just let changes hit us any old way, they simply make us more incoherent, they make us feel bad. But if we join our body-mind into a whole, and open ourselves to take in the extra energy with awareness, then that same energy can raise us up to a new level of energy and insight—we can learn from it rather than feeling bad about it. And it makes a difference to how we respond to events in our life when we realize that we do have the choice to rise up or collapse.

For example, think of a couple who have a relationship—they form a couple-holon. The relationship maintains a steady kind of pattern, or order, so long as they do pretty much the same things and stay pretty much the same (which, of course, humans never do). Inevitably, negative emotions will arise between the couple. The holon can respond to this negativity, which is often very energetic, in two ways. The couple can use the energy of the negativity to reach a new level of understanding and awareness between them, popping the relationship up to a whole new degree of order. Or the couple can just collapse and separate. People generally just do not understand this nowadays. So often marriages end in divorce simply because the couple is unable to see a negativity as an opportunity to rouse themselves.

A similar kind of situation happens in any small group of people, for example, the staff in an office, a spiritual community, or just a group of friends. When negativity arises, as it inevitably does, the group can be open and responsive to each other and to the needs of the group-holon, and reach a new level of working together. Or they can freeze, pull apart, and communicate less effectively from then on.

History shows that whole cultures have collapsed because of their inability to respond to sudden shocks or changes in their environment. Even the United States of America could collapse if it's people continue to believe they are part of a great machine rather than a living organism. And that living organism, is in turn a holon, a part of the larger living organism, which is all of human society on earth. It's not too late, but there is not much time for this change to come about. It may take a big and terrible shock to make it happen.

People who are very awake stir up the energy of the world around them to create shocks that can wake people up to a higher level of inner order. There are many stories of Gurdjieff, the Russian spiritual teacher that I mentioned in my first letter, doing this. And I saw Chögyam Trungpa Rinpoche stir up his environment over and over again. The environment around him could sometimes become so highly irritating that we had just three choices—leave, sink into a gloomy funk, or raise ourselves up to a whole other level of energy, insight, and awareness-feeling. Many people left Rinpoche forever after a brief time with him because they simply couldn't stand the way his presence stirred up their obstacles and irritations. And those of us who did stick around all too often chose the second course—sinking into a brown funk. But sometimes it was possible to raise up our energy, and then what Rinpoche showed us was truly wonderful.

Here is another example of how our holon can adjust to shocks, or a hit of energy. Do you remember I told you in letter 8 about the way our perception makes guesses about what is "out there" to fit with what we expect, in other words to keep our world familiar? Now, we might ask, however, what happens when what we are confronted with just doesn't seem to fit anything familiar in our memory bank?

Our first response to a strange perception is usually a mixture of heightened attention, or wakefulness, and fear. As I told you, I even felt a little uncomfortable when I couldn't make out the pattern on the lake. It was not, of course, that I felt in any way threatened, as if some monster were lying there. But some slight anxiety, along with my awakened curiosity, was an automatic response of my body-mind to the confusion of my visual system at that moment.

A mix of alertness and fear is aroused whenever someone has a perception that doesn't match what she unconsciously anticipates she *should* perceive in a particular situation. Suppose you were sitting alone in the house and you knew that Mom and I left together to go to a movie half an hour ago. Then you hear voices in the next room. You would become very still and your hearing would become very sharp. You would probably be scared for a moment. Then you realize that we left the television on, so you are able to fit a familiar scenario to the voices in the next room.

Trying to keep our perceptions constantly matching our guesses about what is going on is a very tricky business, in which the unfamiliar is always lurking very close by. Our perceptual system usually copes, automatically, by narrowing our lives so that we do not come across the strange too often. Or perhaps we could say that each moment is *actually* unfamiliar but, to avoid fear, we make it familiar by our habitual guesses. In this way, we try to maintain our holon at a steady level of order.

Whenever our perception strays from the familiar, fear pulls us back. So fear keeps us on our habitual psychological track as we go through the world. We could say that fear is the very boundary of our normal world. Fear is the basic energy of the cocoonlike barrier that protects us from experiencing anything strange, which word literally means, "what is beyond the boundary of the familiar."

When we do not recognize the perception process for what it is, our lives are ringed by a circle of fear. And we try, constantly, to avoid experiencing this fear. If, on the other hand, we do recognize the fear and experience it, rather than try to avoid it, then we are not forced to live in the habitual patterns of our upbringing. We can use the shock of the strange to rouse ourselves to a new level of awareness. We can go *into* the strange world with an open attitude of inquisitiveness and curiosity, rather than try to avoid it. So the feeling of strangeness can be a source of tremendous possibility rather than something to pull back from.

This is really the basis of much spiritual effort—being willing to go forward into what makes us scared and awake at the same time, what feels strange, either in the world or in our own state of mind. Instead of pulling back and trying to make our world as small and comfortable as possible, we can take the attitude of fearlessness. We can be willing, even eager, to go forward beyond the boundary of the familiar. And our world becomes bigger and bigger.

# LETTER 21

# *Patterns of Life and Circuits of Thought*

*Dear Vanessa,*

In the previous letter I showed you something about how holons behave when they are interacting with their surroundings—how they can pop up or down to different levels of order when a shock of energy enters them.

Now I want to take us a step further in developing our story of holons. We're trying to overcome our habit of seeing the world as empty, dead space full of things. We're talking about how to see the world as awareness-feeling-energy space full of dynamic patterns of holons. Seeing our world filled, not with things, but with self-organizing energetic patterning holons, can provide us with new insight and feeling for the ordinary things in our world. It can give us a new way of seeing the livingness of things and the thinkingness of things.

First, what makes a holon alive? Well, this is where what I told you in the previous letter about self-organization comes in. What makes a holon alive is its ability to organize itself.

When scientists discovered that the DNA molecule was the basis of genetic inheritance, they jumped to the conclusion that they now had a sure way to tell the difference between a living thing and a nonliving thing: this was the presence or absence of another *thing*, a piece of DNA. With this in mind, biologists, psychologists, sociologists, schoolteachers, and popular science writers have proclaimed that DNA is the key to life.

Some biologists now admit that this simple picture is terribly mislead-ing. How all the parts of a cell are put together is *just* as important as whether DNA is there. If you gently break only the outer membrane of a living cell, which holds it into an organized unit, the cell will cease to be living: it will not be able to divide and multiply. If you destroy the pattern of organization, interrupt the process, it is no longer living.

In spite of all the excitement about DNA in the mid-sixties, scientists still had no idea what made a cell *living*. No one then could make a cell come alive by putting all the parts together in a test tube, even with the magical DNA—and they still can't thirty years later. Rather than being due to the presence of a particular *thing*, such as a DNA molecule, life is *dynamic patterns* of organization and activity.

The degree of internal order in a holon is sometimes referred to by scientists as its coherence. To *cohere* means to stay together. So when a holon is in a more coherent state, it means the parts of it are staying together, or acting together, acting harmoniously. The holon is working more as a whole than as a random bunch of parts. We could say that a living cell is more coherent than a nonliving cell.

A German biochemist, Fritz Popp, has shown some very interesting things about coherence. (Popp, by the way, was one of the faculty at the program I attended at PEAR.) All living cells send out light—this is known as bioluminescence—and scientists can measure the coherence of this light. (I won't try to explain what coherent light is, other than to say that a laser beam is more coherent than the beam from a normal flash-light.) Popp has shown that the coherence of the light a cell emits de-pends on how healthy that cell is—the more healthy a cell is, the more coherent is the light it emits.

For example, the light emitted from a fresh egg laid by a free-range chicken is considerably more coherent than light emitted from a mass-produced egg. The light emitted from cancer cells is clearly *less* coherent than the light from healthy cells. And a dead cell emits no light at all.

Popp asks the question, How do the billions of cells in our body keep in communication with each other so that they can act harmoniously, coherently, when they need to? For instance, as soon as a part of the body is infected, the entire immune system goes into action, almost simultane-ously throughout the body. How do the cells communicate with each other so fast? Popp calculated that the speed of messages along nerve pathways is much too slow. He suggests that light traveling through the cells of our body may be what maintains the coherence of all our cells. And he goes so far as to suggest that, based on this, light may in some way be the carrier of consciousness (we might prefer to say awareness) within our body.

So you see, something fascinating is going on here that relates very directly to us. For it suggests that when our body-mind is in a more coherent state, it is more healthy—not surprising if we are in touch with our body-mind, but interesting to see it demonstrated scientifically.

So the point of all this is that self-organization, the patterning of the holon, is what makes it alive, not the separate bits of stuff in it. And the patterning, or dynamic order, decides not just whether a holon is alive or dead but the quality and intensity of that life.

This gives us a whole new range of possibilities for looking at what is alive. Geophysicist James Lovelock, for example, suggested twenty years ago that the earth's biosphere behaves in many ways like a living organism. He calls this suggestion the Gaia hypothesis, after the Greek earth goddess, Gaia.

The Gaia hypothesis, Lovelock wrote, "is that the entire range of living matter on Earth, from whales to viruses, and from oaks to algae, could be regarded as constituting a single living entity, capable of manipulating the Earth's atmosphere to suit its overall needs and endowed with faculties and powers far beyond those of its constituent parts."

In their book *Life beyond Earth*, Gerald Feinberg and Robert Shapiro go so far as to suggest that the way scientists are searching for extraterrestrial life is quite misplaced. In looking for evidence of life, they say, it is futile merely to look for chemical or biological *things* that are associated with life on earth, such as DNA. It is the patterns of organization and the processes of change that would be indications of life.

Feinberg and Shapiro suggest, quite outrageously for conventional scientists, that such patterns and processes might be found in balloonlike organisms in the atmosphere of Jupiter, plasma life in the interior of stars, radiant life in interstellar gas clouds, solid-hydrogen life on very cold planet surfaces, electromagnetic field energy, magnetic domains in neutron stars, and other extraordinary places.

So we have seen what makes a holon alive, in the ordinary biological sense. Now we can ask what makes a holon capable of perceiving or thinking? Well, obviously a thinking holon has an even greater degree of order and self-organizing activity than a simple cell. And this self-organizing activity may even reach beyond the boundary of the holon. The great biologist and mind scientist Gregory Bateson thought about these things with tremendous freshness, free from the conditioning of his colleagues. He emphasized that if we want to understand any action that involves thinking, it is nonsense to just look at circuits in the brain. We must take into account *complete* circuits, including the "outside."

This morning I was chopping a log for kindling, with an ax. The thinking of my body-mind in this situation must obviously include the *entire* circuit: my brain, my visual system, the muscles of my arm, the ax, the cut in the log. I make each successive stroke of the ax in response to the result of the previous stroke. If I am a bit off center the first time, I correct it the second time. But it has to be a *complete* feedback circuit. We can't draw an imaginary boundary around a *part* of one of these circuits and say *all* of the thinking is in that part. If you block the part of the circuit between my eye and the log, by putting a screen there, I would certainly miss the log and make a horrible mess of my leg.

Now, this is very important; do you see why? It's saying that the circuits that channel my thinking are not restricted just to my body-mind. Bateson suggested that my thoughts form a network of pathways. And this network is not limited to my consciousness or my skin. This network of thought pathways includes the pathways of all my unconscious mental processing, and we saw how extensive that is in my letters about perception. But these pathways also include all *external* tracks along which information travels so that I can chop the log. The part of the circuit from eye to log must be included in the thought pathway just as much as the part of the circuit from eye to brain.

Perhaps this all seems quite obvious to you, but it took someone of the stature of Bateson to see it and point it out for us. Nevertheless, mainstream scientists have taken little notice of this aspect of Bateson's work, just as they took little notice of Whitehead's work on feeling. Whitehead and Bateson simply didn't go along with the "mind is in the brain" delusion that has so powerfully ensnared the minds of most scientists.

So what does all this add up to? When we realize that mental processes, such as perception or thought, are patterns of self-organization, then we can look for these patterns anywhere. There is no reason at all for the patterns to be restricted to the brain. In fact, as Bateson says, it doesn't make any sense to localize my thought in my brain when I am chopping the log. Nor do the patterns of thought have to be found only on the scale of ordinary human life. Bateson himself examined the patterning of the way species have changed over the course of millions of years and suggested that this patterning is just like the patterning of thought in an individual. So, he suggested, the whole process of evolution could be regarded as a process of *thought*. Bateson wrote, "The total self-corrective unit which processes information, or 'thinks' and 'acts' and 'decides' is a *system* [holon] whose boundaries do not at all coincide with the boundaries either of the body or of what is popularly called the 'self' or 'consciousness'; and it is important to notice that there are multiple differences between the thinking system and the self as popularly conceived [for example, isolated in the brain]. . . . The individual mind," he wrote, "is immanent not only in the body. It is immanent also in the pathways and messages outside the body; and there is a larger Mind of which the individual mind is only a subsystem."

But we can let our visualization roam even beyond this. Thoughtlike patterns could exist at scales as small as a single cell, or even the genetic material, or at the level of quantum waves and the geometry of space. Thoughtlike patterns could also exist on a galactic or cosmic scale. Feinberg and Shapiro suggested some pretty novel patterns that could be the basis for life. Couldn't all these patternings just as easily be complex enough to be circuits of thought? Sure! Why not?

The estimated number of stars in the Milky Way galaxy is almost identical to the estimated number of neurons in the human cortex. The possibilities for exchange of energy between galactic systems are vast and there is amazing order in galaxies. Why could a galaxy not be a circuit of awareness? And what about cosmic patternings that are too large for us to detect? As Paul Davis, professor of physics at the University of Adelaide, says, "Nature is a product of its own technology, and the universe is a mind. Our own minds could then be viewed as localized 'islands' of consciousness in a sea of mind."

Where are your circuits of awareness located? And where are the circuits of awareness located that you are part of ? Are they somewhere in your head? Are they throughout your whole body? Or are they part of a system larger even than this? The holon that is your brain is part of the holon that is your body. And this is part of the holon of your family, your society, your local ecosystem, your whole world. Each of these holons provides pathways for awareness that you can tune in to when your heart, your organ of feeling, awakens.

Perhaps thinking about the simple example of me chopping a log can help you to see what I have been saying all along in these letters about feeling-awareness reaching out in webs of tentacles to touch what seems outside of us. And what seems outside of us, too, can reach back and respond to us with tentacles of feeling.

When communities perform elaborate dances or rituals, such as the Javanese shadow theater, the Navaho Holy Ways, or even the Super Bowl, there is often a feeling of awareness in the whole pattern of activity.

This is why ritual, ceremony, and celebration are so valuable, in fact necessary for the health of the community—they provide individuals with ways of entering into a level of awareness that they can more easily access through the activity of the group.

And what of energies that are too subtle for us to see, or hear, but that nevertheless we can feel ourselves to be part of? What of the patternings that we feel as dralas, gods, angels, and so on. Aren't all these circuits of thought and feeling gathering-places for awareness? There's a lot of room for drala energy patterns here, don't you think, Vanessa?

Sometimes you can try to look at your world with a new vision. A lot of the time we see it as just full of lumpy things. But sometimes you can try to see it with your heart, to feel it full of patterns or whirlpools of life, feeling, and thinking in a wholeness that, too, is aware and full with feeling. And you, your body-mind, your feeling, your awareness, are part of those whirlpools. A new vision: try it, you'll like it.

# LETTER 22

❦

# *Third Interlude*

*Dear Vanessa,*

For this third interlude I will write about another meditation practice that I do in the mornings after a period of sitting. The practice is connected with the principle of resonance and "as above, so below," which I will be writing quite a lot about in our last few letters.

If you place a tuning fork on a grand piano and strike the appropriate key, the tuning fork starts to vibrate. It's tuning to the larger energy of the piano and being affected and energized by it, but the note it sounds is its own. Likewise, if we place our body-mind in the appropriate way, we can invite larger energies and forces in the cosmos to resonate through us and awaken our own energy and wisdom. This particular practice that I am going to describe invokes, or rouses, a particular kind of awareness-feeling-energy, that of Manjushri.

Let me explain first what, or who, Manjushri is. Manjushri is one of the great beings of the Buddhist spiritual tradition, who vows to help all beings. He does not have a material body, like us, but we imagine him in human form (I write "he," but there are similar beings of feminine gender). He is a kind of benevolent and powerful energy-awareness running throughout the whole universe. The energy Manjushri embodies is the energy-awareness of intellect, insight, and intuition all joined, or heart and mind joined. In other words, Manjushri's awareness is how we know about our world when we are a whole person: we know intellectually and at the same time we know intuitively.

With the knowing of Manjushri we know something through and through. We know it with our sharp intellect, but we also know it so deeply that, in a way, we become it. Perhaps the closest way I can illustrate this for you is by describing loving someone. If you love someone, not necessarily sexually but in the way that a mother loves her child or you might love your best friend, you give some part of yourself to that person. You open to her completely. And because you are so open, you see without judgment all the details of how she is. But beyond seeing this, you know something deeper about her, you know *who* she is, and who she is not, in a deep way. This sounds a bit idealized, I know, but it might give you some idea of the way we can know with Manjushri's insight-energy.

Another way of describing the way Manjushri knows is to say that he sees just as Kathleen Raine said she saw, in the quote in my first letter: he sees that everything is connected to everything else by being part of one whole, undivided world. Within this whole, everything stands out more clearly as itself than it ever does when you see the world as just many separate things; yet you see each thing as not having separate *thing-ness* at all. It's a bit like when you watch a movie and you see separate people, loving each other, shooting each other, and so on. But you know that they are all part of one movie, one light shining through a projector onto one screen.

This is the best form of knowing and it is very much related to the feeling-awareness kind of knowing that I have been writing about—how we know the roses or each other when our chattering minds are quiet. Do you remember I wrote about participating consciousness (which is much the same as what I have been calling feeling-awareness) when I was writing about the alchemists of the Middle Ages? Well, Manjushri knows with participating consciousness as well as with conceptual knowledge. This may sound a bit complicated and wordy, Vanessa, but if you feel it at the same time as you think about it, it's quite simple.

Manjushri's skin is white but translucent, with a yellow glow like the sky at dawn. He sits cross-legged, and is ornamented with the robes and headdress of an Indian prince. He has a gentle and youthful smile. In his left hand he holds a cup of amrita, a liquor symbolizing intuitive insight. In his right hand he carries a two-edged sword, symbolizing the sharpness and precision of intellect. One edge of this sword cuts through all our confusion, doubt, and egoism that prevent us from knowing in the whole way that he does. And the other edge of the sword cuts through any pride that might develop because you *have* the sword—because that pride would be just another obstacle to knowing like Manjushri.

So, back to the practice of invoking Manjushri. The practice is very traditional and yet very simple. We always begin with sitting practice. Into the space and openness that develops in sitting practice we let go of our usual sense of self, good ole everyday *me*. We let go of everything—thoughts, feelings, hopes, and fears. And within that same open space we imagine, visualize, feel, that Manjushri arises. He does not have flesh and bones, like everyday me. He is simply a body of light arising like a hologram from the energy-feeling-awareness of space. And we visualize-imagine-feel that we are Manjushri. We imagine that we are simply a hollow, light body, sitting like Manjushri, dressed like Manjushri, smiling like Manjushri, holding Manjushri's sword, *feeling* like Manjushri, and softly chanting the sound of Manjushri—OM, ARAPACHANA, DHIH, HUM. This sets up a kind of vibration, just like striking that piano key. If we do it clearly and precisely, it can attract the energy of Manjushri that fills all of space.

This may seem a bit more complex than sitting practice, perhaps even strange. But I think you can probably see that mindfulness and awareness are key elements of this practice, just as they are in every aspect of our life—here, mindfulness is paying careful and precise attention to the details of how Manjushri looks, feels, and sounds; awareness is opening, and extending into fathomless, formless energy of Manjushri.

(By the way, this can be a powerful transformative practice. I would not recommend that you try it from this description, or without proper preparation and instruction. I know you wouldn't do this, Vanessa, but I write this for anyone else who reads these letters. There is a great deal more to the Manjushri practice than I have described here. I have been practicing Manjushri for only twenty years, and I am very much a beginner.)

There are many practices of inviting deity energies, and awakening their energy and insight within us, like the Manjushri practice. In Buddhism there are practices for awakening compassion or fearlessness. And there are practices for developing the ability to pacify situations that need pacifying, to enrich situations that need enriching, and to destroy obstacles to compassion and wakefulness when they need destroying. Navajo rites of healing and initiation, called the Holy Ways, are another example.

And of course there are similar dances and rites in many native traditions, as well as the Shambhala teachings.

The instructions for practices like the Manjushri practice emphasize that the practitioners need to be as precise as they can about the visualization. Presumably this is necessary to enable the resonances to arise. Did you ever see the old TV advertisements for Memorex tape in which the great jazz singer Ella Fitzgerald broke a wine glass by singing a note, and the commentator asked, "Is it real or is it Memorex?" Ella had to sing the resonance tone of the glass absolutely precisely in order to cause it to vibrate enough to break, and of course the point of the ad was that Memorex tape could be that precise as well. Practices like the Manjushri practice are somewhat like that.

Practices of visualization and sound repetition, such as the Manjushri sadhana, do actually feel very like "tuning in" to a radio station—you have to find the spot, and when you've found it, you enter a different feeling of energy and awareness; you tune in to the energy-feeling quality of the deity. When you understand and feel how Manjushri arises out of space, you can understand how everything in our so-called ordinary world arises in just the same way—like holograms out of the fullness of space.

And these practices have to be done not just with the outer form, the bodily motions and so on, but with the appropriate inner attitude of attention and feeling. To do them we need to synchronize our body, feeling, and awareness. But at the same time, doing the practice can itself *help* us to synchronize our body, feeling, and awareness precisely because it sets up resonances between the three that can help us to tune them together into the greater energy space.

At the end of the Manjushri practice session, Manjushri dissolves, and almost immediately our ordinary experience of "me, myself" is there again. At that moment we can realize that, just like every other so-called *thing* in the world, this "me, myself" is not so solid as we usually tend to think it is. I am not so much a solid, unchanging *thing* as a dynamic, ever changing pattern of energy, feeling, and awareness, just like Manjushri. That split second, between Manjushri dissolving and our familiar self arising, can be a moment to experience *now*.

When our familiar self dissolves momentarily, there sometimes remains an almost overpowering feeling of joy and sadness mixed. We feel joy because we can finally let go of the thing that has been depressing us most of all, our heaviest burden, and the deepest seat of our conditioning—our "me." And we can let go of the awful struggle that has been constantly with us all our life, the struggle to satisfy me, to make me happy, wealthy, successful, famous, brilliant, and even enlightened. At the same time, we feel a kind of bittersweet sadness, realizing that our only companion, the one who has been with us all our life, the only one who really knows us, our close and intimate friend, has been fooling us all along—into thinking we *are* "me," only me, and nothing but me.

Some practices, such as the mindfulness-awareness type of practice, emphasize seeing our "me," and seeing through it. Other practices, such as the Manjushri type, show us the awakened energy that is present, and that we are, beyond the me.

Mindfulness-awareness practices, or practices of silence, and Manjushri-type practices, or practices invoking the energies of particular deities, are complementary to each other. Practices of silence settle the busy, chattering thoughts and emotions, so that we can begin to see through the tight veil, or cocoon, that thoughts and emotions put up between us and our world. Practices like Manjushri connect us directly to universal energy that is awake, beyond language and conditioning.

But all practices work with that boundary between holding on and letting go. They all recognize that at the moment of letting go, a miracle can happen, the miracle of *nowness*. So, here's just one more slogan:

OPENING, TOUCHING, RESONATING-FEELING,
LETTING GO—*nowness!*

# LETTER 23

❦

# *Joining Heaven, Earth, and Human*

*Dear Vanessa,*

In yesterday's letter I wrote about how self-organizing holons gather energy to raise themselves to higher levels of order. I wrote about holons as patterns of life and thought from the very smallest scale to the scale of galaxies. And I suggested that these patternings of life and thought could be the channels of awareness-feeling with which we connect with others and our environment. How we make this connection is through what I have been calling resonance—again, just as a second guitar string resonates when the string on the first guitar is plucked.

Now, to understand this principle of resonances better, I want us to look at another quality of the patterning in holons. Many holons are built up through a single pattern, repeating over and over at different levels of scale, from the very small to the very large. The single pattern that repeats to produce a large holon is itself a lower-level holon, of course, because everything in the universe is a holon! (Just a reminder.)

Patterns appearing at one level of magnitude repeat themselves again and again for as much as millions of size changes. And through this repetition of patterning, we are able to tune in to the awareness-feeling-energy ocean on many different levels.

When you look at a cloud, it seems to have a definite shape that often seems to remain unchanged for a fairly long period of time. Yet we've all seen those magnificent movies of clouds speeded up. The clouds appear

— 185 —

seething in the most beautiful regular and irregular dynamic patterns. And scientists have found the same patterns when they zoom in to look at smaller and smaller sections of the cloud, down to a millionth the size of the original cloud.

Patterns that repeat themselves over and over at different scales were given the name *fractals* by Benoit Mandelbrot in 1975. Mandelbrot's work has dramatically changed the way we think of the forms in nature.

Mandelbrot pointed out that a tiny piece of coastline still looks like a coastline if it is magnified ten or a hundred times. The pattern of the coastline looks very much the same if you take an aerial view of a large section or a detailed map of a smaller section, or even go down to a rock-by-rock outline of one hundred yards of coast or a pebble-by-pebble map of one foot. He said that a coastline is highly "self-similar."

A coastline is just like a cloud in this respect, and like innumerable other formations in nature. This property of self-similarity of patterning at many orders of scale is perhaps one of the most significant and far-reaching characteristics of nature. It is a profound principle of the way holons are built up from smaller holons. It is quite possibly how much of nature forms itself. Remember the five-skandha patterning of experience in letter 9?

Fractals have revealed a new way of looking at the endless levels of beautiful intricacy in nature. Clouds, ferns, the root or branch systems of a tree, the nervous system and the system of veins and arteries, the tributaries of a river, the pattern of a flash of lightening—all have this quality.

Mandelbrot discovered some fractal objects that are truly marvelous. To gaze at them relaxes the eye and opens the mind. These amazing forms are made by iterating (repeating over and over) a very simple mathematical formula on a computer. When these iterations are done millions of times and the results plotted on a graph, the most beautiful and intricate two-dimensional shapes can be seen. And the shapes are repeated as you go deeper and deeper into them.

One of these extraordinary objects is known as the Mandelbrot set. The basic Mandelbrot set is seen in figure 8. It looks a bit like a gingerbread man. We can zoom in on a small section of it, as shown in figure 9. Zooming in again, we get figure 10, again, and we get figure 11. Magnifying more than a million times, we find figure 12, with a perfect gingerbread man in it! The process could go on forever. And we could have started off with a Mandelbrot set a billion or a trillion times bigger. The Mandelbrot set is just one of infinitely many ways to generate these delightful repeating patterns.

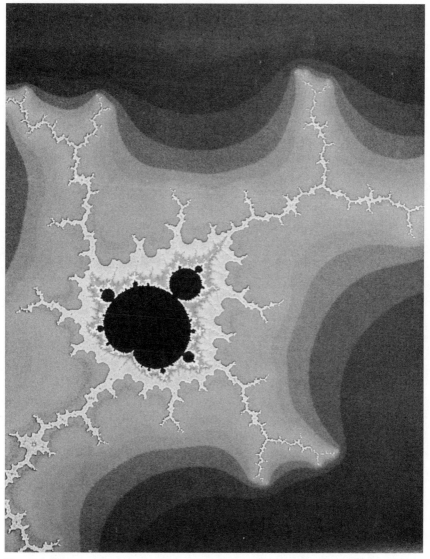

FIGURE 8

Fractals have also helped scientists to describe the complex patterns of *dynamic* order within nature. The way the behavior of some holons changes over time is often so complex as to appear chaotic—totally lacking in order. Yet scientists are now finding order deep within systems, such as the weather, that had appeared to be completely chaotic.

Think, for example, of weather patterns or the smoke from a cigarette

FIGURE 9

sitting on the edge of an ashtray, the pattern of someone's emotions over a period of time or the patterns in the history of an organization or a whole nation. All may seem at first to be quite chaotic. The smoke from the cigarette curls upward in an intricate, often fascinating, and almost hypnotic form. The form is constantly changing, never the same for a moment, yet there is some self-similarity in it. Something about these patterns seems to stay the same as the smoke curls up. Is there any way

FIGURE 10

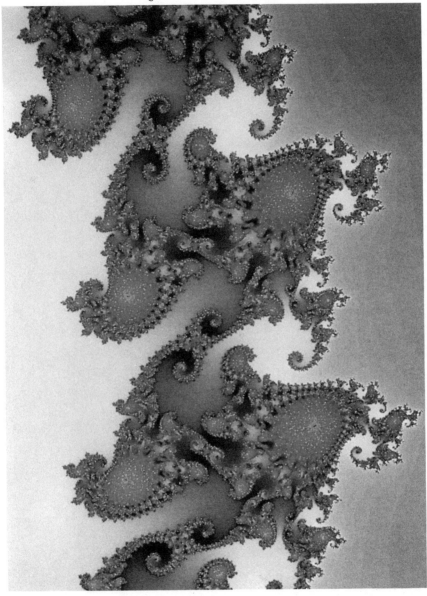

FIGURE II

of describing the apparent order within the ever changing chaos of the smoke?

Scientists have a way of defining the overall state of a complex holon, a weather pattern, for example, as one point on a special kind of multi-

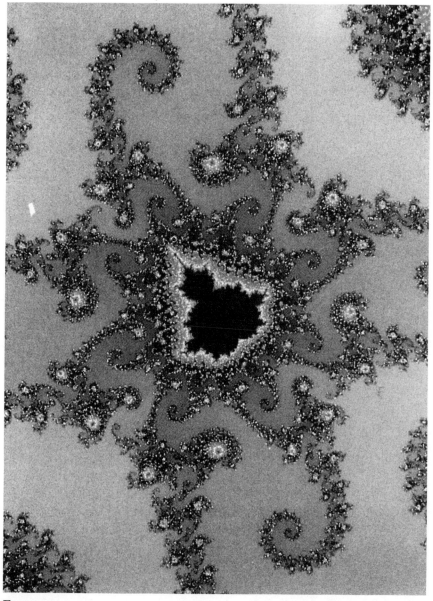

FIGURE 12

dimensional graph, called the *phase space* of the holon. They plot the position of this point as it moves through time. Just as with the Mandelbrot sets, these multidimensional graphs became possible only with the advent of supercomputers. I won't try to say what the phase space is, but the point right now is that *one* point on the graph represents the dynamic state of the complete holon at a particular point in time. So as time goes along, the movement of the point in phase space tells something about how the holon, as a whole, is changing.

When the behavior of a holon such as the weather is plotted in phase space, the most extraordinary and visually beautiful creatures appear (figure 13). These patterns, which show the order hidden deep within chaos, are called *strange attractors*.

And this brings us back to our patterns within patterns. The strange attractors have one special characteristic—they are fractals. Strange attractors never repeat—the holon can go on indefinitely and it will never be quite the same again. The cigarette smoke or the weather, even our emotions, never quite go through the same pattern twice. Yet *something* about their dynamic state is self-similar: a similar pattern is found no matter at what scale we examine the strange attractor. So there is a fascinating kind of order hidden deep within even the most chaotic-seeming systems.

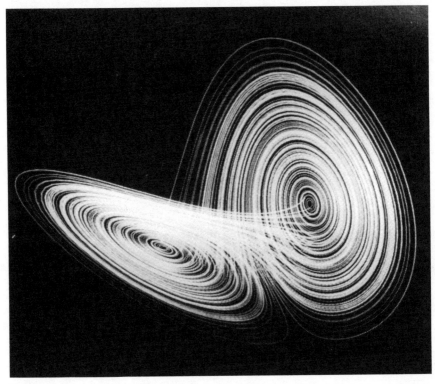

FIGURE 13

Fractals and strange attractors are not *objects* that we can measure. And it is hard to even analyze them, or grasp them logically. But we can feel the pattern, and realize that patterns like this are found, repeating slightly differently over and over again, in virtually every dynamic holon in nature. So, although the weather or the cigarette smoke or sometimes our thoughts and emotions may *seem* chaotic, there is deep order hidden in all of them.

These dynamic fractal patterns are at the heart of meaning and quality. The difference between a plastic flower and a living flower lies in the complexity of its structure and the deep levels of order hidden within. When we have a feeling for pattern, when we connect patterns on many levels, we can almost literally "see infinity in a grain of sand."

Feeling-energy-awareness would be completely formless if it were not for all the patterns that spontaneously play in it, from the tiniest of tiny to the biggest of huge. It's the patterns that provide the qualities of the variety of different things in our world, the variety of holons. By feeling the patterns we feel the quality, and we can feel the quality repeating at different levels.

This gives us a way to understand how it is that we can feel, resonate with, patterns of energy-awareness-feeling greater than our own scale—gods, angels, dralas, and all the rest. Feeling a pattern of energy vibrating in our own body-mind, we can resonate with a similar pattern at any magnitude. This is what the medieval alchemists perhaps meant by "as above, so below." And a sixth-century Ch'an (Chinese Zen) Buddhist, Seng-ts'an, wrote, "The very small is as the very large when boundaries are forgotten; the very large is as the very small when its outline is not seen."

Most spiritual practices involve repetition. This is the basis of most forms of living ritual: the repetition of mantras, creative visualization of deities, repetitive ritual drumming and dancing in native traditions, and so on. By repeating a sound or creating an image over and over again, we are tuning in to a particular energy-awareness-feeling pattern at the level of our ordinary body-mind. We thereby evoke the resonance of similar patterns at other levels.

It's just like our tuning fork analogy. The tuning fork communicates with a grand piano—if a piano tuner strikes a tuning fork, let's say sounding the note middle C, and holds it on the sounding board of the piano, then if the middle C string of the piano is tuned correctly it will sound. The grand piano will resonate in response to the tuning fork! It is perhaps in a similar way that humans can communicate with dralas. We create in our own being a pattern that resonates with theirs—as above, so below—and so can connect with or tune in to their energy pattern.

A wonderful example of this principle is a story of a Chinese rain-maker. It was told to Carl Jung by Richard Wilhelm, who lived in China for many years and was the first translator of the *I Ching*. His is still the classic translation that most people use. Jung told his biographer and colleague, Barbara Hannah, to tell this story in every lecture. I quote Marie-Louise von Franz:

> In Kiaochou came a great drought so that men and animals died in the hundreds. In despair, the citizens called for an old rainmaker who lived in the mountains nearby. Richard Wilhelm saw how the rainmaker was brought into town in a sedan chair, a tiny little gray-bearded man. He asked to be left alone outside the town in a little hut, and after three days it rained, and even snowed! Richard Wilhelm succeeded in being allowed to interview the old man and asked him how he made the rain. But he answered, "I haven't made the rain, of course not." And then, after a pause, he added, "You see it was like this—throughout the drought the whole of nature and all the men and women here were deeply disturbed. They were no longer in Tao. When I arrived here I became also disturbed. It was so bad that it took me three days to bring myself again into order." And then he added, with a smile, "Then naturally it rained."

The Western traditions of alchemy and astrology (*real* astrology, not those silly horoscopes in the newspaper we sometime giggle at), as well as the Confucian and Taoist traditions, recognized correspondences between various realms of existence: inanimate, plant, animal, human, and heavenly, or perhaps we would say material, biological, psychological, and spiritual. In fact, these correspondences formed the very basis of the spiritual understanding in these systems.

Paracelsus, for example, who was one of the great alchemists, wrote that all the patterns of heaven and earth are to be found in humans. In any philosophy or medicine, he said, the macrocosm—the entire cosmos—must be found in the microcosm, the individual person.

Here is a similar idea, by Jacota Rai, a healer from Bali, Indonesia: "If you want to be strong and healthy, listen to your body. Learn from your life and don't allow negative emotions to manifest in your body. It's not the individual health you work for. Your body is a microcosm of the earth itself. Take care of your body and you take care of the earth."

Through correspondences between heaven and earth and humans, events that are apparently unconnected from one view become intercon-nected when a broader viewpoint is taken. What appear to be "chance" coincidences when we consider only a narrow realm of experience are recognized as connected when a larger realm is taken into account. They are connected through the self-similarity of their patterning.

An example of this principle is the Snake Dance, described by Trudy Sable in her thesis on the Mi'kmaw peoples' traditional ways of transmit-ting knowledge to their children. In the Snake Dance a line of dancers, often children, snake around the dance floor with their hands on each other's shoulders or waists. The dance is accompanied by the sound of chanting and the rattling of a horn filled with pebbles.

At one point the leader stops and the line spirals in around him to form a tight coil, and then reverses and spirals out again. The line moves across to the other side of the dance floor and the spiraling in and out is repeated. The form of the dance is like the form of a medicinal plant, called the *meteteskewey* . This, too, curves in a spiral and has seeds that rattle, like a rattlesnake. The medicine extracted from the *meteteskewey* can also be poison if it is not gathered at the proper time and used in the proper way. So the plant has a protector who needs to be properly hon-ored. The protector of the plant is a serpent, or snake, the *jipika'm*. There are legends and stories about all these elements—the plant, the medicine, the turning of the seasons, and the protector serpent. This is what Trudy Sable says about the Snake Dance:

> In a sense, there are three levels of meaning to the dance, all insepara-ble. On the external level, the rattling of the horn, filled with pebbles, mimicked or reflected the rattling or tapping sound of the plant. The chant that accompanied the dance, Margaret Johnson suggested, may have mimicked the beat of the rattling leaves or stalk—*metetesk, met-etesk, metetesk*—onomatopoetically. This plant, in turn, would be pow-erful medicine for the people if properly respected. . . .
>
> On another level, the dance and chant most likely was a part of becoming, awakening, honoring, and possibly testing the energy of the *jipika'm*, the essence of the medicinal plant. The plant itself most likely mirrored the features of the *jipika'm* in both appearance and the sound it made. The sound and rhythm of the dance embodied the essence or nature of the *jipika'm*, which was inseparable from the medicine. . . .
>
> The third level of the dance may have to do with "turning over" the seasons, and also be connected in some way with the stars. At this level the dance would be performed to mark or effect the changing of the seasons. . . . There are two particular legends which talk about the turning over of the seasons, in connection with medicines.

Trudy suggests that this dance could be taught in an elementary science curriculum, and concludes:

This dance is yet another teaching of respect for the powers at play, some of which can kill you, such as picking the wrong medicine. It also teaches of the seasons, the directions, the stars, the nature of reptiles, the bird that leads one to the medicine, and values of respect and care needed in collecting plants. Offerings to the four directions were made in the dance, acknowledging the gift of medicine. Properly approaching the medicine will be good and strengthen the people. From the study of one dance or one plant, a whole web of relationships and information about the world comes into being.

In the Greek tragedies, the realms of existence larger than human were symbolized by gods who could at times enter into and affect the affairs of humans, while conducting their own affairs independently of the human realm. Such times are turning points in the lives of the people who are touched by the gods. I say "symbolized" but I am sure that for the Greeks their gods actually did enter into their affairs at turning points like those depicted in the plays. Their plays were a slice of their lives, just as "Seinfeld" or "One Life to Live" or "Beverly Hills 90210" are slices of life for our culture. Take your choice!

A very similar situation occurs in Javanese shadow theater, the wayang. These plays are performed for whole villages. They usually happen at night and can go on for several nights in a row. It is assumed that not only the living human villagers attend but also ancestors, gods, and all manner of spirits and demons. The gods and spirits are actually considered the essential audience, while the human audience is just passing through.

The language of the plays incorporates all levels of the history of Javanese language—from Old Javanese and Sanskrit (in which the gods are addressed) to modern Javanese and American slang. The performance of the play is an auspicious meeting of gods and men. This is reflected in the plots of the plays themselves, which are built primarily around auspicious or meaningful coincidence.

Within the play, as within life, there are various worlds and various times occurring simultaneously and occasionally interweaving. And when the worlds do meet, that is a moment of coincidence with powerful meaning and consequences.

The real subtlety and teaching of the wayang appears in this coincidence of worlds. In a single wayang there is the world of daimons—the direct sensual world of raw nature. There is the world of the ancestor heroes. There is the world of the ancient gods—a distant cosmological world of pure power. And there is the world of the clowns—a modern pragmatic world of personal human survival. All of these realities coexist and between each of them there may be confrontations, meetings, battles. Each exists in a different concept of time, and all the times are occurring simultaneously.

Alton Becker, an anthropologist who lived in Java for many years, writes, "Nature time, ancestor time, god time, and the present are all equally relevant in an event, though for each the scope of an event is different. Shadow theater, like any live art, presents a vision of the world and one's place in it which is whole and hale, where meaning is possible. In all its multiplicity of meaning a well-performed wayang is a vision of sanity."

And let's not forget that for the Javanese, these worlds *actually* meet. It is not just a fairy story. For that matter, nor were fairy stories just fairy stories in our world, once upon a time.

Well, Vanessa, in this letter I've introduced you to the levels of patterning that permeate reality through and through—the dynamic fractals of nature, life, and mind; the strange attractors, which show these dynamic fractals at work even within the midst of apparent chaos, bringing pools of order within that chaos that connect multiple levels of magnitude. And I've shown how, in many different traditions, this is expressed in the principle "as above, so below," the principle of microcosm and macrocosm—that the human reflects and is reflected in heaven and earth.

This afternoon I want to suggest that these things are not just for other traditions and cultures, but we too can connect and resonate with many levels of meaning in our own lives by feeling the auspicious coincidence of each moment, by appreciating *nowness*.

# LETTER 24

❦

## Being a Tuning Fork on a Cosmic Piano

*Dear Vanessa,*

I wrote about auspicious, or meaningful, coincidence in some of my earliest letters. And so, appropriately, for time is not linear after all, we come full circle to look at coincidence again.

You go to the store and meet a friend whom you haven't seen for five years and did not know was in town. You happened to be thinking about your friend that very morning because you rather urgently needed to get some information that only he had. This is a meaningful coincidence. Or your computer breaks down just when you are about to write a letter that will get you in hot water, which happened to me some years ago. But can we be open to the larger meaning, the many worlds that are coming together in that coincidence, just as worlds meet in the Greek plays or the Javanese wayang? Can we let our lives be guided by this larger vision, as it shows itself to us in coincidence?

There have been clear instances of meaningful coincidence while I've been in this retreat place, writing this book, Vanessa. I told you about one of them in letter 5—the coincidence of the hot water running out on the very morning that I needed to be reminded of coincidence. Let me tell you about another that happened just this morning. I was going over what I had written a few days ago, about how holons could either jump to a higher level of energy or crash to a lower level with less order in response to a shock. I'd been fussing with this for several days. I had originally written it in a rather stiff style, and I couldn't see quite where it fit in. And I felt that it was very much in need of some practical example.

So this morning I was rewriting that passage to try to make it simpler and clearer. And as I was writing, I was wondering, "What would be a good example?" At that very moment I heard an unfamiliar noise: bah-*gom*, bong, bong; bah*gom*, bong, bong. . . . I stopped typing to listen—it was the faint sound of music. As I listened it got louder and I recognized some kind of light country rock. The caretaker's house is across the drive-way, about two hundred yards from my retreat cabin, and I thought, "My goodness, can Mark really be playing music that loud?" Didn't seem like his kind of music, though.

So I went to the door and looked out. There, parked in the driveway between my cabin and Mark's house, was an old, dark green pickup. Someone was in the driver's seat, holding a cup of Tim Horton's coffee, head bopping in time to the music.

My first thought was to go and ask him, "Please, could you turn the music down, it *is* a bit disturbing." But then I had the thought, "This is *exactly* what I'm writing about." So I went back to my seat at the computer, feeling the irritation and letting it energize me. It worked; after a few moments I didn't mind the music at all, especially since it seemed to be there for the explicit purpose of helping me with this difficult passage. Before I had finished the passage, the music was turned down, and within a few minutes the truck drove off.

Coincidence? Yes, of course. And it was *meaningful*. There was a coincidence of my outer world and the inner world of my thoughts about this book, which is very intense, because I am eating, sleeping, and practicing Manjushri with it.

There have been many incidents like that. I came here to write in retreat because I was having quite a bit of difficulty finishing this book. I have felt that Manjushri has been very close and I have asked, sometimes quite passionately, for his help. Actually, that is how I came to be writing these letters to you. I had written two rather dense, long, and stiff books on the topic of science and spirituality ten years ago. Mom had encouraged me for a long time to try to write a simple version that everyone could understand.

After struggling for two years with trying to write that simplified version, I decided to give it one last chance, go on retreat for three weeks, and if nothing happened—forget it forever! Before I went on retreat I had been particularly concerned about the way the dumbness of school was affecting you. And I had also been strongly affected by a conversation with a young friend, Adam, who was in the first year of college. He told me that many of his close friends from high school were in a desperate situation—they were clinically depressed, on hard drugs, or, in at least one case, had committed suicide. He wanted to impress on me how terribly despondent his, and your, generation are at the state of the world and the hopelessness of what they have in their future.

On the first morning of my retreat, I sat practicing Manjushri and found myself saying, "Please *help*." And I heard, or perhaps I should say felt, physically in my chest, the words, "Dear Vanessa." I thought, "Oh, that's nice, but it won't work." Then again, I heard, "Dear Vanessa," and again I dismissed it. When I heard "Dear Vanessa" a third time, I got up and started writing, "Dear Vanessa . . . ," and the book flowed from that.

As you know, I have written three books before this one, and writing has always been rather a struggle for me. It has been rare that more than a few sentences have flowed easily off my fingers. And it has almost always felt like good ole plodding Jeremy putting down his good ole familiar thoughts. But during this past three weeks there have been times when the words flowed faster than I could type them, and when it felt as if someone were typing through me.

I tell you this story not because I think my experience is particularly special, not at all, in fact, quite the contrary. I am suggesting that *anyone* can connect with their drala friends, which they may feel simply as an intense intuition, if they have the passion and openness to do it. And if their vision is not obscured by their belief that the rational scientific world story is the only real story of their world. But you have to have longing, and you have to be willing to love. The great Sufi, mystic, poet Rumi, lover of the divine within and without, said, "If you want to develop a special organ of perception, O necessitous one, increase your necessity!"

Lucia Roncalli, a tremendously dedicated and profoundly caring midwife, whom I met at the PEAR conference in 1994, tells of meeting her personal drala while attending at a home birth. Describing this fourth encounter, she says,

Actually I am not just feeling it inside me. I am seeing it in the room. It has looked the same each time I have seen it, and I call it by proper name now: Dread. It is a huge mound, somewhat resembling a mountain and somewhat resembling a buffalo. It has thick curly hair, and an eye—I see it in profile. It is not personally threatening, nor even monstrous. It simply is, and is with us, another presence in the room.

Dread is gendered—I think of it as male, though to call Dread "him" personifies overmuch. The other three times Dread has appeared, normal births from healthy women who have had normal pregnancies have careened off the tracks. The emergencies have been unforseeable and bizarre. The outcomes have been good, but getting there has involved interminable moments on the razor edge between life and death.

On this fourth occasion that Lucia had felt Dread—in two hundred successful home births she had attended—the baby's vital signs were healthy all the way up to seconds before the birth. Yet the baby survived only a few months after birth, on a ventilator. She died when her parents finally decided to pull the plug. Should Lucia have insisted that the birth take place in hospital, against the strong wish of the mother and father, and against the advice of her midwife partner—even though there was no clinical reason to do this? Lucia comments, "The thing about funny feelings is—in a rationalistic culture you can be talked out of paying attention to them. And if you respect "hard data" and the "hard data" contradict feeling, then you enter Coyote territory: a subtle and unchartable realm in which cookbook rules do not apply."

Lucia concludes:

Possibly [the baby's] life could have been saved if I had accepted the information of Dread and suggested we transport [the mother to the hospital.] Or possibly there was another factor at play, one her parents could have understood and translated if I had been open with them about Dread. They may not have opted for transport and the hospital; they may have opted for prayer or something else. Then, at least, the decision about what to do and the consequences would have belonged, as they should, fully to the parents.

These possibilities still haunt me and make me want to bang pots and ring bells and shout from the rooftops: *Intuitive knowledge is just as real as laboratory data!!! It needs to be included in clinical decision-making with just as much respect as what can be quantified! We need to develop a language to talk about these things, and talk about them well!*

Peoples in almost all societies throughout the ages have invoked such energy patterns—gods, spirits, helpers, dralas, angels, devas, whatever we call them. In the cosmic story that I am telling, the "gods" are simply patterns in the energy-awareness-feeling ocean that we can tune in to, and draw out, or invoke. We are not ultimately separate from the deities we invoke, and by participating in their being, we can communicate with them and they can communicate with us.

When we allow the dralas into our system, even for a moment (which may not last longer than a blink!), they can guide us. And how we begin to feel these connections, and to let the dralas in, is only through living *now*. Nowness is the key; nowness is the only moment you have, actually, Vanessa. Now, nowness, this moment, that's it!

But I want to warn you of one thing here, Vanessa. The energy patterns of gods and spirits are not *necessarily* pleasant, or helpful, so we'd better watch out! There are energy patterns at the scale of dralas that can divide the world and cause suffering, just as there are at the human scale. We don't want to invite those! We can tune in to larger patterns of feeling-energy to benefit ourselves and others, but we can also do a lot of harm. It's like electricity, neutral, and can be used for great benefit or great harm. We should know who we are calling on, know what we are doing and why we want to do it. The best safeguard is to understand that *ultimately* we are not and cannot be divided one from another. Ultimately we are profoundly connected with each other in feeling-energy-awareness space. So when we want to invoke drala energy, we should always think and feel that we are doing it not just for ourselves but always for ourselves and others.

The rituals that connect us to a big world, an enchanted or sacred world, are *not just spiritual practices*. And I really want to emphasize this! Everything we do, chopping wood, lighting a fire, cooking breakfast, starting the car, walking down the produce aisle of a supermarket, all are also rituals. We repeat these day after day in our lives. And by being present when we do them, we can begin to feel how each of these activities connects with everything else in our world. Crafts, such as wood carving, or any activities that humans have practiced over and over, including sports, can connect us with the specific drala energies of those activities. By doing the craft in accordance with its tradition, and with real passion, we repeat the patterns that have been passed down through the generations and have the power to attract dralas.

My whole life is made up of activities that I repeat over and over. And when I can feel the patterns of what I do, rather than merely narrowing my focus to the activity in front of my nose (nowness is not narrow but open; awareness is there as well as mindfulness), then my life becomes a connected whole. And the energy level of that whole is very different from times when my life is a scattered series of unconnected and unfelt zombie actions.

So, feeling the small actions of our life, putting our heart into them, and feeling the patterns they form in our life can raise our life up. Then we begin to tune in to the patterns that go beyond our small world and connect us into a much bigger world, the sacred world that includes the dralas and gods.

When we connect into the patterns of the bigger world, we find an ocean of energy available to us. It's very simple, but we do have to stop blocking ourselves from the awareness-feeling-energy ocean of which we are a part, stop resisting and pulling away in fear. When we jump into this ocean of energy, when we feel a part of it, a playful exchange can begin to happen. Our need to struggle for this or that dissolves and we can relax. Things come to us and we respond, rather than react. We turn and whirl, lean in and lean out, give out and take in, and circle around. We can feel the humorous contrast to our usual serious solidity and can appreciate that things happen on time.

I am not talking about being happy all the time, or constantly having "fun." You may feel quite depressed, but if you really let yourself feel that, go into it, taste it, you can begin to feel the energy and wisdom even in that depression; you can feel the depression as part of the play, part of the larger picture.

We can bring about profound change in our culture by tuning in to this ocean of energy and the drala patterns of our individual worlds and our communal worlds. We are all connected to the larger world, and to other worlds, drala worlds, in any case. But if we constantly deny it, as our culture does today, then we will continue to feel cut off and alone in the universe. And we will continue to feel depressed and helpless.

There is great suffering in our world, all around us, and we must try to help, genuinely help. But no matter how much social or political action we engage in, or how much spiritual practice we do, thinking we are helping ourselves and others, we will only create further depression if we still deeply, and unconsciously, experience a Dead World. Our actions will unknowingly perpetuate and promote the Dead World, which is the *real* cause of depression.

People say, "I'm fed up with hearing about vision; when are we going to get practical?" But it's a question of priorities; if you don't change your vision, at the depth of your being, changes in pragmatics will make no difference. When your vision of society changes, then practical solutions *do* emerge naturally.

We are all on the earth together, in a world we mutually create. We are all holons within a society, which is a holon within the world. And the world at the moment is in a dangerous mess. We won't survive as isolated individuals disconnected from each other. Individualism goes along with the Dead World. By tuning ourselves and helping others to tune in to the enchanted world, our actions will naturally follow our vision and truly help to bring about a good society. And it will be a good society not just for humans but for dogs and trees and dralas as well.

As a community, we can resonate together like tuning forks on the cosmic piano, to raise up our power and energy. And by power I mean our ability to connect with the world—not our ability to make someone else do what we want.

Our society is used to thinking we can solve our problems by having the controlling, manipulating kind of power over the earth and everything on, under, and above it, rather than by touching and being with the earth. It hasn't worked, and I think you and your friends know this, Vanessa, and feel it very strongly.

One way to touch the earth is by celebrating it. Celebration and ritual, collaborating as a community, are crucial ways for people to gather together their life-giving connections with their world. Through ritual, people celebrate harmony with the world, gather their power together collectively, and call on their dralas to celebrate life with them. Rituals set up resonances so an energy exchange between dralas and the human community can happen. We connect with the elemental forces in the world. Connecting with the water element and water dralas can connect with the water element in ourselves and the earth. Celebrating the wind element tunes us in to the wind and the swirling energy of the sky. When feeling or love energy is sent out along energy-awareness pathways, magical results can happen.

In our Western society, celebration has become a way to escape from the drudgery of life rather than a way to connect with life and earth. Parties are empty events when people put on polite masks, or when they are "blowouts" that deaden body/mind/heart. Attempts at ritual are often tedious and boring. It wasn't always this way, though. The pre-Christian "pagans" had elaborate rituals and formalized festivals at major turning points in the cycles of nature.

Then these festivals were turned into holidays like Christmas, Easter, All Saint's Day, and so on. Sacred stone circles were smashed, sacred wells were filled in, sacred trees were cut down, sacred people were burned. People were forbidden their life-giving rituals. When the Puritans got involved, the demons were finally driven out by not allowing any singing (except hymns), dancing, playing, or sensuous activity whatsoever. You were just supposed to keep your nose to the grindstone, work hard, and keep away from the Devil. If you let yourself relax or get out of control in any way, the punishment was severe. We're still living in the shade of this Puritan legacy. People are still caught up in individualism and dreary workoholism.

But is it so far-fetched, after all we've been talking about, to think that rain dance ritual is more powerful than damming lakes or seeding clouds with pollutants? Does putting people in prison create less violence than helping them experience their own goodness and gentleness? Is constant busyness more effective than dancing with the dralas and inviting their help? Ordinary solutions are helpful in their limited way, but perhaps we can indeed expand our thinking.

Celebration and ritual are important parts of life and are as necessary as food. Ceremonies are the observance of natural order that align humans with natural events and connect them with the natural spirits. Spirits are involved and come alive in the ritual. As the ancient Chinese said, nature creates ceremonies and humans observe them. Celebration is for life itself. It's a natural overflowing of alive joy, a renewal, a revitalization. Feeling the goodness and delight in oneself extends to feeling the goodness of others and the boundless universe we're all part of.

# LETTER 25

Our *Never-Ending Story*

*Dear Vanessa,*

Well, Vanessa, in these last few letters, we have had a small taste of a different way of viewing the things of our world. We've come a long way, and in this final letter I would like to summarize what all this might mean for you.

Remember that all along we have been talking about stories. None of this is any kind of absolute truth. But the stories we tell each other make our common world. And especially the stories we tell our children profoundly influence them. The stories they hear make the world they grow up into, and the world they will make when they are adults.

You are about to enter into the world as an adult, and you could make a real difference in it. You could help others to live in a sacred, living world. That is why I have written these letters, in the hope that you and some of your friends will read them. Please stay clear of the Dead World, Vanessa. And please help other people into the world of the living, a world that I know you know.

I showed you the way our perception creates our world. And I showed you how the beliefs that we grew up with, as well as our interpretations about the world and our automatic emotional reactions to the world, enter into that creation without our being aware of it. Because of this, the story you grew up with profoundly affected your perception and your world. In letter 3 I briefly told that story for you.

I have also shown you how much of that story was *made up* by scientists, or by popular writers in the name of science. But I have also tried to demonstrate for you in these letters that the story you grew up with is not the *only* one that science could tell. It is a fabrication, a fantasy story, made up by people using science for purposes other than pure science itself.

We could say that in these letters I *too* have made use of science for a purpose other than science itself. But the way I have used science is very different from the way it has been used in the past. In the past, and even most often now, science was used to try to convince us to *believe* something. What I have wanted to do in these letters is to use science to *cancel* the beliefs that conditioned you as you grew up. I have tried to help you to *unbelieve*, or debelieve, rather than to try to get you to believe even more stuff. I wanted to help you to free your perception, the perception of the heart, from the bonds that those beliefs trapped you in.

There is a difference, though, between the *other* story that science can tell, the one I have been telling in these letters, and the story we grew up with. The other story acknowledges *awake feeling*, the perception of the heart. And the other story acknowledges what is seen with that perception, the ocean of awareness-feeling-energy and the patterns that form in it. So we could, perhaps, say that although the other story too is just a story, it is a story told with the senses and heart a little wider open.

And there is an important *scientific* reason for preferring the other story over the old story. In science, when there are two theories that both seem to explain the same set of observed phenomena, one way to decide between them is to say, "Well, which of the theories can include the widest range of phenomena in its explanations?"

I think it should be clear to you now that the *other* story can certainly accommodate everything that conventional scientists accept as real phenomena. But it can also accommodate, and give a coherent, rational, and intelligent explanation for, a wide range of phenomena that are constantly being observed but that conventional scientists won't even look at. I will give you some examples. First, of course, is all the work on psychokinesis and precognitive remote viewing that has been done at PEAR and many other laboratories, which I wrote about in letter 17.

Then there are other events that people report so often that they simply *must* be included in any reasonable story. For example, another phenomenon that conventional science has no way of dealing with is out-of-body experiences, in which people feel that their awareness leaves their body and travels to another location. In some cases, people are able to gather information about the world that later checks out. Here are three of the many cases that have been reported, always by people who seem to have no reason to make them up:

Dr. Josef Issels, a German cancer specialist, reported:

> One day I experienced a remarkable happening. I was doing my morn- ing round on Ward One, the ward reserved for the acutely ill. I went into the room of an elderly woman patient close to death. She looked at me and said, "Doctor, do you know that I can leave my body?" I knew approaching death often produced the most unusual phenomena. "I will give you proof," she said, 'here and now." There was a moment's silence, then she spoke again, "Doctor, if you go to Room 12, you will find a woman writing a letter to her husband. She has just completed the first page. I've just seen her do it." She went on to describe in minute detail what she had just "seen." I hurried to Room 12, at the end of the ward. The scene inside was exactly as the woman had de- scribed it, even down to the contents of the letter. I went back to the elderly woman to seek an explanation. In the time I had gone she had died."

The second story is told by a woman who was driving back from Lon- don to Luxembourg, where she worked with the European Parliament:

> About midnight I hit a patch of black ice going round a curve on the edge of a ravine and went into a skid. My left tire hit a low barrier separating the two lanes of traffic which caused the tire to burst. The car turned over and last thing I remember is feeling ice on my face. I next found myself spiraling out of my body, rushing back over the route I had come. I noticed back around the bend in the road an on- coming car with an elderly couple inside. The man lifted his hands off the wheel when he heard the crash, at the same time the pipe he was smoking dropped into his lap. He turned to his wife and said, "Here we go again." I saw him trying to brush the cinders from the fallen pipe off his trousers. I then felt myself being drawn up a dark tube or funnel. There was a vague impression of a figure who offered me a beautiful shawl, all white and soft and luminous. I heard him say, "Come, you must be so cold, so tired.". . .
>
> I remember saying, "No, I'm sorry, I can't accept your lovely shawl as I'm all covered in blood." I looked down and could see myself inside the car covered in blood. I don't remember how I returned. The next thing that happened was I regained consciousness. Everyone was amazed for they were convinced I was dead. The next morning I said

to the [elderly] man, "When you heard the accident, why did you say 'again'?" He said, "I don't understand." So I said, "You turned to Madam and at the same time dropped your pipe out of your mouth." He turned extremely white and said, "But Madam, you couldn't possibly know that because you were in the crash around the corner."

Social worker Kimberly Clark, of the University of Seattle, tells of a patient called Maria, who suffered a cardiac arrest while in the hospital where Kimberly worked:

Maria said to Kimberly, "The strangest thing happened when the doctors and nurses were working on me: I found myself looking down from the ceiling at them working on my body."

I was not impressed at first. I thought that she might know what had been going on in the room, what people were wearing, and who would be there, since she had seen them all day prior to her cardiac arrest.

Then Maria proceeded to describe being distracted by an object on the third floor ledge on the north end of the building. She "thought her way" up there and found herself "eyeball to shoelace" with a tennis shoe, which she asked me to try to find for her. She needed someone else to know that the tennis shoe was really there to validate her out-of-body experience.

With mixed emotions I went outside and looked up at the ledges but could not see much at all. I went up to the third floor and began going in and out of patients' rooms and looking out their windows, which were so narrow that I had to press my face to the screen just to see the ledge at all. Finally, I found a room where I pressed my face to the glass and saw a tennis shoe! My vantage point was very different from what Maria's had to have been for her to notice that the little toe had worn a place in the shoe and that the lace was stuck under the heel and other details about the side of the shoe not visible to me. The only way she would have had such a perspective was if she had been floating right outside and at very close range to the tennis shoe. I retrieved the shoe and brought it back to Maria; it was very concrete evidence for me.

It is not particularly difficult to accept these stories if we understand that awareness can extend beyond the body, as we do in the *other* story. Of course, there is a great deal of interesting research that could be done to explain just how these things happen.

An important phenomenon that has been reported frequently and investigated carefully is distance healing. At the Mind Sciences Foundation, in San Antonio, Texas, a series of experiments were done in which someone in one room made an effort to influence the state of calm or arousal of a volunteer in another room. As of 1991, they had held 323 sessions with 271 different volunteers. It was very clear in these tests that one person's physiological/emotional state was being influenced at a distance by the intention of someone else.

Related to this, there is the phenomenon of healing by group prayer, which has also been clearly demonstrated in many different circumstances. For example, Dr. Randolph Byrd, a cardiologist, studied almost four hundred patients who had had heart attacks. The patients were divided into two groups. Both groups were given state-of-the-art medical care. The difference between the two groups was that one group was prayed for as well. The prayed-for group did so much better than the other group that if a drug had had the same effect, pharmaceutical companies would have been panting to get their hands on it.

Distance healing, too, is a phenomenon that we could begin to understand in the *other* story, but it has to be ignored, or even denounced by conventional science, simply because it just does not fit in that old story!

Now I would like to ask you to look back for a moment to letter 3 and revisit the story I told you there, the story you grew up with.

OK, now let's contemplate the *other* story. As I have shown you through these letters, the other story could just as well be the story that science tells, if only scientists themselves had not been so conditioned by the old story as they grew up, just as you were. Scientists are humans and subject to exactly the same conditioning, and blindness in regard to that conditioning, as we are.

Here, then, is the *other* story that science could tell, and the story that I have tried to tell you in these letters:

We live in a profoundly good world, which is permeated through and through by feeling, caring, awareness. We live in a huge space, so huge it is beyond imagination, huge and unfathomably deep. That space, from deepest depth to the very surface that we see and hear and smell, is feeling, energy, awareness. We, and everything we see and taste and touch, are part of that space. That space is in our body, in our minds. Better to say *we* are in that *space*. We are like holograms arising playfully in that space.

We can enjoy that living world. Feel the joy of being able to leap into living space even if only for a moment. And feel the joy of at least knowing and trusting it's presence here, now, so that we can fundamentally relax. Feel the joy of *being* space—feeling, living, aware space. That is what we want to do, and what we can do. We can relax, because we are filled through and through with good, gentle, fiercely caring space. And through that benevolent space we are profoundly connected to everything else. Everything else, too, is a playful living surface in our feeling, caring space.

That space is playful because it is not manipulative; there is no Master Planner controlling us from a separate Heaven. The world of energy-awareness-feeling space is a purely spontaneous creative show. Knowing this, we do not have to constantly try to grab and hold on to every little thing. We can let things come to us and respond to them, just as children spontaneously play in the sunshine or run for shelter in a thunderstorm.

Many people are in pain, though, because they do not realize that they can play in living space. They feel they are stuck, dead lumps in a dead, empty world. This is a great sadness. It is our own sadness as well, because we too are in pain when we forget that we are living space. And we often forget this, so we are often in pain. Our joy can never forget this sadness. So the feeling of awake heart is always joy and sadness together.

The sadness is our own sadness, too, because the sadness of others *is* our own sadness. And when our heart perception is open we *cannot* forget this sadness. So we need to help others to play. That could be the best purpose in life—to learn how to feel living space and to play there. And then to bring others along to play there as well.

We can see-feel our connection to all else that rises up playfully in space through perception of the heart. Seeing with the heart, hearing with the heart, smelling with the heart, tasting with the heart, touching with the heart, and feeling with the heart. This way we connect ourselves to others. We connect along the webs of feeling that cross through us and stretch way out into space, and way *in* into space.

Sernyi, Mom, you, me, your friend Margaret, the rock on our hill, the trees in Point Pleasant Park, the red barn outside my kitchen window, the car with a flat tire, the watch that keeps stopping. All these shine with their own luminosity, their own energy, their own space. All these glow when we cherish them with feeling-awareness.

We can know things when we put our feelings out to them. Your car that won't start—put your feeling out to it, you can *feel* why it won't start. The roses—when I put my feeling out to them, I *know* what they need. Sernyi looking with *that* look—when you put your feeling out to her, you know what she wants to tell you. Put your feeling-awareness out to the earth—you can feel its nourishment and support and know its grieving for its own dying. Put your awareness-feeling out to space and discover wisdom-insight beyond the thinking mind.

Some people say they talk to flowers. Other people say, "Nonsense, flowers don't talk, and what language do you think they are speaking anyway? English? Ha, ha, ha." No, flowers don't speak in English, stupid. Flowers speak in *feeling*, through patterns of energy. And when we look, but really *look*, at a flower, or anything else for that matter, we can feel their pattern of feeling. We can resonate with it. The words people use to talk to flowers are the only way they know to communicate feeling, but that is what is happening—feeling is going out to the flowers. And feeling is resonating back from the flowers. Our world is a glorious un-ending circling out and back of feeling-energy-awareness. Generosity.

But don't let's limit our world too much. There may be more to it than meets the eye. Spirit helpers, angels, gods, dralas, call them what we will, perhaps we can know these, too, when we learn to open our heart, when we learn feeling-perception. And we can invite them into our life. We can resonate with many levels of patterning and connect with the great ocean of energy-feeling-awareness by paying attention to the small details of our life and how they all connect together. So, let's celebrate the vastness, the livingness, the goodness of the world! Let's invite Sernyi and the rock and the gods (and the demons too, if they dare) and anyone else who would like to join us. That could be a beginning of a good society.

❦

Well, Vanessa, tomorrow this retreat ends—it's back home for me. It will be strange but cheerful to see you and Mom after I have spent three weeks writing letters to you. And back to the office—see those old famil-iar patterns!

I have tried in this last letter to summarize the journey that we have been through together in these past weeks. For, although you have not actually been in the cabin with me, you have very much been here in feeling, and I do feel that we have taken a journey together.

All my love to you, Vanessa, and to your generation.

Your Dad

# POSTSCRIPT

# *References and Further Reading*

GENERAL

There are quite a lot of books that deal with science and spirituality. Many speak of science and spirituality as if they were two objective realities, like two parallel worlds. They rarely touch on the important question of how we know our world—the sciences of mind such as neuroscience and cognitive psychology—nor do they seem to understand the role that Dead World science plays in our own deep conditioning. Though they struggle to bring science and spirituality together, many seem to be still trapped in the body-mind-nature split. The following books try, like these letters, to take a broad perspective that sees scientific knowledge and spiritual insight as complementary ways of knowing one world:

Harman, Willis. *Global Mind Change: The New Age Revolution in the Way We Think*. New York: Warner, 1988.

Hayward, Jeremy. *Perceiving Ordinary Magic: Science and Intuitive Wisdom*. Boston: Shambhala Publications, 1984.

Needleman, Jacob. *A Sense of the Cosmos*. New York: Doubleday, 1975.

Peat, F. David. *The Philosopher's Stone: Chaos, Synchronicity, and the Hidden Order of the World*. New York: Bantam, 1991.

Varela, Francisco, Evan Thompson, and Eleanor Rosch. *The Embodied Mind*. Boston: M.I.T. Press, 1992.

Wheatley, Margaret. *Leadership and the New Science.* San Francisco: Berrret-Koehler, 1992. Not about spirituality as such, but a delightful illustration of how an understanding of the new story of science can lead to an altogether different approach to leadership in business.

Wilber, Ken *Sex, Ecology, Spirituality.* Boston: Shambhala Publications, 1995. This book is daunting—nearly eight hundred pages, of which more than two hundred are notes. But it *is* brilliant in Ken's inimitable way: witty, erudite, insightful, and impatient.

LETTER 1: THE LIVING WORLD WE FELT AS CHILDREN

Gurdjief, G. I. *All and Everything.* New York: E. P. Dutton, 1964.

Jeans, James. *The Mysterious Universe.* Cambridge: Cambridge University Press, 1931.

Moore, James. *Gurdjieff.* London: Element Books. 1991.

Ouspensky, P.D. *In Search of the Miraculous.* New York: Harcourt Brace Jovanovich, 1965.

Raine, Kathleen. Quoted in *Facing the World with Soul,* by Robert Sardello. New York, Harper Perennial, 1994.

Speath, Kathleen Riordan. *The Gurdjieff Work.* Los Angeles: Tarcher, 1989.

Trungpa, Chögyam. *Born in Tibet.* Boston: Shambhala, 1977.

Webb, James. *The Harmonious Circle.* Boston: Shambhala, 1987.

Wilber, Ken. *Quantum Questions.* Boulder: New Science Library, 1984.

LETTER 2: STORIES WITH FEELING, STORIES WITH SOUL

Arden, Harvey. *Dreamkeepers: A Spirit-Journey into Aboriginal Australia.* New York: HarperCollins, 1994.

Black Elk, Wallace and William S. Lyon. *Black Elk, The Sacred Ways of a Lakota.* New York: HarperCollins, 1991.

Bloom, Harold. *Omens of Millenium: The Gnosis of Angels, Dreams, and Resurrection.* New York: Riverhead, 1996.

Boyd, Doug. *Rolling Thunder.* New York: Delta, 1974.

Campbell, Joseph. *The Power of Myth.* New York: Doubleday, 1988

Erickson, Carolly. *The Medieval Vision.* New York: Oxford University Press, 1976.

Evans-Wentz, W.Y. *The Fairy-Faith in Celtic Countries.* Oxford: Clarendon Press, 1911.

Fox, Matthew and Rupert Sheldrake. *The Physics of Angels: Exploring the Realm where Science and Spirit Meet.* New York: HarperCollins, 1996.

Gersi, Douchan. *Faces in the Smoke.* Los Angeles: Tarcher, 1991.

Gold, Peter. *The Circle of the Spirit: Navajo and Tibetan Sacred Wisdom.* Rochester, Vt.: Inner Traditions, 1994.

Harpur, Paul. *Daimonic Reality: Understanding Otherworld Encounters.* New York: Arkana, 1995.

Hayward, Jeremy. *Sacred World: A Guide to Shambhala Warriorship in Daily Life.* New York: Bantam, 1995.

Heinze, Ruth-Inge. *Shamans of the 20th Century.* New York: Irvington, 1991.

Isozaki, Arata. *"Ma: Space-Time in Japan,"* Japan Today: 36

Knudtson, Peter, and David Suzuki. *The Wisdom of the Elders.* Toronto: Stoddart, 1992.

Lawlor, Robert. *Voices of the First Day: Awakening in the Aboriginal Dreamtime.* Rochester, Vt.: Inner Traditions, 1991.

Maclean, Dorothy. *To Hear the Angels Sing.* Hudson, N.Y.: Lindisfarne Press, 1990.

Maybury-Lewis, David. *Millenium: Tribal Wisdom and the Modern World.* New York: Viking, 1992.

McGaa, Ed, Eagle Man. *Mother Earth Spirituality.* New York: Harper-Collins, 1989.

McNeley, J.K. *Holy Wind in Navajo Philosophy.* Tucson, Az.: The University of Arizona Press, 1981.

Moore, Thomas. *Care of the Soul.* New York: HarperCollins, 1992.

Morgan, Marlo. *Mutant Message.* Lees Summit: Mo., MM Co., 1991.

Ono, Sokyo. *Shinto: The Kami Way.* Rutland, Vt.: Charles Tuttle, 1962.

Parisen, Maria, ed. *Angels and Mortals, Their Co-Creative Power.* Wheaton, Ill.: Quest Books, 1990.

Sheridan, Joe. Tells the story about the Cree writer Stanley Wilson in the chapter "Sacred Land—Sacred Stories: The Territorial Dimension of Intuition" in Robbie Davis-Floyd and P. Sven Arvidson, eds., *Intuition: The Inside Story: Interdisciplinary Stories and Perspectives.* New York: Routledge, 1996.

Thundup Rinpoche, Tulku. *Hidden Teachings of Tibet.* London: Wisdom, 1986.

Trungpa, Chögyam. *Shambhala: The Sacred Path of the Warrior,* Boston: Shambhala, 1984.

van der Post, Laurens. *The Lost World of the Kalahari*. New York: Harcourt, Brace Jovanovich, 1986.

van Ness Seymour, Tryntje. *When the Rainbow Touches Down*. Seattle: University of Washington Press, 1989.


Letter 3: The Story of the Dead World

Crick, Frances. *The Astonishing Hypothesis: The Scientific Search for the Soul*. New York: Touchstone, 1994.

Delbruck, Max. *Mind from Matter?* Oxford: Blackwell, 1986.

Humphrey, Nicholas. *Leaps of Faith: Science, Miracles, and the Search for Supernatural Consolation*. New York: Basic Books, 1996.

Monod, Jacques. *Chance and Necessity*. New York: Vintage, 1971.

Peacocke, Arthur. *Intimations of Reality*. Indiana: Notre Dame U.P., 1983.


Letter 4: How Our World Was Disenchanted

Achterberg, Jeanne. *Woman As Healer*. Boston: Shambhala, 1990.

Berman, Morris. *The Reenchantment of the World*. Ithaca: Cornell University Press, 1981.

Brehier, Emile. *The History of Philosophy, Volume V: The Eighteenth Century*. Chicago: University of Chicago Press, 1967.

Burke, James. *The Day the Universe Changed*. Boston: Little, Brown, 1985.

Butterfield, Herbert. *The Origins of Modern Science*. New York: Free Press, 1965.

Collingwood, R. G. *The Idea of Nature*. Oxford: Oxford University Press, 1960.

Eisler, Raine. *The Chalice and the Blade*. New York: Harper and Row, 1987.

Goldstein, Thomas. *Dawn of Modern Science*. Boston: Houghton Mifflin, 1980.

Heer, F. *The Medieval World*. New York: New American Library/Mentor, 1964.

Pagels, Elaine. *Adam, Eve, and the Serpent*. New York: Vintage, 1989.

Shepherd, Linda Jean. *Lifting the Veil: The Feminine Face of Science*. Boston: Shambhala, 1993.

Singer, Charles. *A Short History of Scientific Ideas*. Oxford: Oxford University Press, 1959.

Szamosi, Geza. *The Twin Dimensions: Inventing Time and Space.* New York: McGraw-Hill, 1987.

Thomas, Keith. *Religion and the Decline of Magic.* London: Penguin, 1971.

Whitehead, Alfred North. *Science and the Modern World.* New York: Free Press, 1967.

Whitrow, G. l. *The Natural Philosophy of Time.* Oxford: Clarendon Press, 1980.

Yates, Frances. *Giordano Bruno and the Hermetic Tradition.* Chicago: University of Chicago Press, 1964.


## LETTER 5: THE ENCHANTED WORLD IS *Now*

Coveney, Peter, and Roger Highfield. *The Arrow of Time: A Voyage through Science to Solve Time's Greatest Mystery.* New York: Ballantine, 1990.

Denbigh, Kenneth. *Three Concepts of Time.* Heidelberg: Springer-Verlag, 1981.

don Jose, Matsuwa, in Joan Halifax, *Shamanic Voices.* New York: Arkana, 1979.

Fraser, J.T. *Time the Familiar Stranger.* Amherst: Massachussetts U.P., 1987.

Koestler, Arthur. *The Roots of Coincidence.* New York: Vintage, 1972.

Ornstein, Robert, and Paul Ehrlich. *New World, New Mind: Moving Toward Conscious Evolution.* New York: Touchstone, 1989.

Priestley, J. B. *Man and Time.* London: Aldus Books, 1964.

Shallis, Michael. *On Time.* New York: Schocken, 1983.


## LETTER 6: FIRST INTERLUDE

Borysenko, Joan. *Minding the Body, Mending the Mind.* New York: Bantam, 1987.

Chödrön, Pema. *The Wisdom of No Escape and the Path of Loving-Kindness.* Boston: Shambhala, 1991.

Epstein, Mark. *Thoughts Without a Thinker.* New York: Basic Books, 1995.

Goldstein, Joseph, and Jack Kornfield. *Seeking the Heart of Wisdom.* Boston: Shambhala, 1987.

Goleman, Daniel. *The Varieties of Meditative Experience.* New York: E. P. Dutton, 1977.

Hayward, Jeremy. *Sacred World: a Guide to Shambhala Warriorship in Daily Life.* New York: Bantam, 1994.

James, William. *Psychology: A Brief Course.* New York: Dover, 1961.

Kabat-Zinn, Jon. *Full Catastrophe Living.* New York: Delacorte, 1990.

Kornfield, Jack. *A Path with Heart.* New York: Bantam Books, 1993.

Polanyi, Michael. *Personal Knowledge.* Chicago: University of Chicago Press, 1962.

Thrangu, Khenchen. *The Practice of Tranquility and Insight.* Boston: Shambhala, 1993.

Trungpa, Chögyam. *The Path is the Goal: A Basic Handbook of Buddhist Meditation.* Boston: Shambhala, 1995.

## LETTER 7: DOES YOUR BRAIN SEE?

Calvin, W. *Cerebral Symphony: Seashore Reflections on the Structure of Consciousness.* New York: Bantam, 1990.

D'Amasio, Antonio. *Descartes' Error.* New York: Avon, 1994.

Eccles, John, and Karl Popper. *The Self and Its Brain.* New York: Springer International, 1981.

Edelman, Gerald. *Bright Air, Brilliant Fire: On the Matter of the Mind.* New York: Basic Books, 1992.

Globus, Gordan. *The Postmodern Brain.* Amsterdam: John Benjamin, 1995.

Harth, Erich. *Windows on the Mind, Reflections on the Physical Basis of Consciousness.* New York: Quill, 1983.

Hayward, Jeremy, and Francisco Varela, eds. *Gentle Bridges: Conversations with the Dalai Lama on the Sciences of Mind.* Boston: Shambhala, 1992.

Hooper, Judith, and Dick Teresi. *The 3-Pound Universe.* New York: Macmillan, 1986.

Laughlin, Charles, John McManus, and Eugene d'Aquili. *Brain, Symbol, Experience.* New York: Columbia U.P., 1992.

Penfield, Wilder. *The Mystery of the Mind.* Princeton: Princeton University Press, 1975.

Poppel, Ernst. *Mindworks: Time and Conscious Experience.* New York: Harcourt, Brace, Jovanovich, 1988.

Searle, John. *Minds, Brains, and Science.* Cambridge: Harvard U.P., 1984.

Weiskrantz, L, et al. "Blindsight, " *The Lancet.* April 1974.

## LETTER 8: INTERPRETATION COLORS OUR WORLD

Bruner, Jerome S. *Beyond the Information Given: Studies in the Psychology of Knowing.* J. Anglin, ed. New York: W W. Norton, 1973.

Dixon, Norman. *Preconscious Processing.* Chichester: John Wiley and Sons, 1981

Flanagan, Owen. *The Science of the Mind.* Cambridge: MIT Press, 1984.

Gardner, Howard. *The New Sciences of Mind.* New York: Basic Books, 1985.

Goleman, Daniel. *Emotional Intelligence.* New York: Bantam, 1995.

Gombrich, E. H. *Art and Illusion.* Princeton: Bollingen, 1969.

Gregory, Richard. "Visual Perception and Illusions," in Jonathan Miller, *States of Mind.* New York: Pantheon, 1983.

Gregory, Richard. *The Intelligent Eye.* New York: McGraw Hill, 1970.

Johnson, Mark. *The Body in the Mind: The Bodily Basis of Meaning, Imagination, and Reason.* Chicago: Chicago U.P., 1987.

Konner, Melvin. *The Tangled Wing.* New York: Holt, Rinehart, and Winston, 1982.

Mandler, G. *Mind and Body: the Psychology of Emotion and Stress.* New York: Norton, 1984.

Miller, Jonathon, ed. *States of Mind.* New York: Pantheon, 1983.

Rivlin, Robert, and Karen Gravelle. *Deciphering the Senses.* New York: Simon and Schuster, 1984.

Seligman, Martin. *Learned Optimism.* New York: Knopf, 1991.

Wilding, John M. *Perception: from Sense to Object.* London: Hutchinson, 1982.

LETTER 9: THE CREATIVE DANCE IN A PERSON

Conze, Edward. *Buddhist Thought in India.* Ann Arbor: University of Michigan Press, 1970.

Guenther, Herbert V. *From Reductionism to Creativity: rDzogs-chen and the New Sciences of Mind.* Boston: Shambhala, 1989.

Hayward, Jeremy. *Shifting Worlds, Changing Minds: Where the Sciences and Buddhism Meet.* Boston: Shambhala, 1987.

LaBerge, Stephen. *Lucid Dreaming.* New York: Ballantine, 1985.

Trungpa, Chögyam. *Glimpses of Abhidharma.* Boston: Shambhala, 1987.

Varela, Francisco. "Living Ways of Sense-Making." In *Disorder and Order,* edited by Paisley Livingston. Stanford: Anma Libn, 1984.

———, Evan Thompson, and Eleanor Rosch. *The Embodied Mind.* Boston: M.I.T. Press, 1992.

Wilber, Ken, Jack Engler, and Daniel Brown. *Transformations of Consciousness.* Boston: New Science Library, 1986.

## Letter 10: Language and a Feeling for the World

Abram, David. *The Spell of the Sensuous: Perception and Language in a More-than-Human World*. New York: Pantheon, 1996.

Barfield, Owen. *Saving the Appearances*. New York: Harcourt, Brace, Jovanovich, 1965.

Bruner, Jerome S. *Actual Minds, Possible Worlds*. Cambridge: Harvard University Press, 1986.

Chalmers, A . F. *What is This Thing Called Science?* Queensland: University of Queensland Press, 1982.

Goodman Nelson. *Ways of Worldmaking*. Indianapolis: Hackett, 1978.

Kosko, Bart. *Fuzzy Thinking*. New York: Hyperion, 1993.

Kuhn, Thomas. *The Structure of Scientific Revolutions*. Chicago: University of Chicago Press, 1962.

Lakoff, George and Mark Johnson. *Metaphors We Live By*. Chicago: University of Chicago Press, 1980.

Lyons, John. *Language and Linguistics*. Cambridge: Cambridge University Press, 1981.

Rorty, Richard. *Philosophy and the Mirror of Nature*. Princeton: Princeton University Press, 1979.

Sable,Trudy. *Another Look in the Mirror*. Halifax: Saint Mary's University MA Thesis, 1996.

Sacks, Oliver. *The Man Who Mistook His Wife for a Hat*. New York: Harper Perennial, 1990.

Suppe, Frederick. *The Structure of Scientific Theories*. Champaign: University of Illinois Press, 1974.

Wallace, B. Allan. *Choosing Reality*. Boston: Shambhala, 1989.

Whitehead, Ruth Holmes. *Six Worlds: Stories from Micmac Legends*. Halifax: Nimbus, 1983.

Whorf, Benjamin. *Language, Thought and Reality*. Cambridge: MIT Press, 1956.


## Letter 11: Awakening Feeling

Csikzentmihalyi, Mihaly. *Flow: The Psychology of Optimal Experience*. New York: Harper and Row, 1990.

Mathieu, W.A. *The Listening Book*. Boston: Shambhala, 1991.

McNiff, Shaun. *Earth Angels: Engaging the Sacred in Everyday Things*. Boston: Shambhala, 1995.

Pert, Candace. In *Healing and the Mind,* by Bill Moyers. New York: Doubleday, 1993.

Sardello, Robert. *Facing the World with Soul.* New York: HarperPerennial, 1994.

Sardello, Robert. *Love and the Soul: Creating a Future for the Earth.* New York: HarperCollins, 1995.

Seng-ts'an. "On Trust in the Heart." In *Buddhist Texts through the Ages,* edited by E. Conze. New York: Harper, 1964.

Williamson, Marianne. *A Return to Love: A Reflection on the Principles of A Course in Miracles.* New York: Harper Collins, 1992.

LETTER 12: NATURE, A GLORIOUS GAME OF COOPERATION. SURPRISE!

Bonner, John Tyler. *The Evolution of Culture in Animals.* Princeton: Princeton University Press, 1980.

Crook, John. *The Evolution of Human Consciousness.* Oxford: Clarendon Press, 1983.

Darwin, Charles. *The Origin of Species.* New York: Mentor, 1958.

Dawkins, Richard. *The Selfish Gene.* Oxford: Oxford University Press, 1976.

Farrington, Benjamin. *What Darwin Really Said.* New York: Schocken Books, 1982.

Grasse, Pierre. *Evolution of Living Organisms.* New York: Academic Press, 1977.

Griffin, Donald. *Animal Thinking.* Cambridge: Harvard University Press, 1984.

Huxley, Thomas. "The Struggle for Existence in Human Society." In *The Nineteenth Century,* February, 1888.

Kohn, Alfie. *The Brighter Side of Human Nature.* New York: Basic Books, 1990.

Kropotkin, Petr. *Mutual Aid.* London: Porter Sargent, 1954.

LaBorde, Roger. In *Sacred World,* by Jeremy Hayward. New York: Bantam, 1995.

Masson, Jeffrey Moussaieff. *When Elephants Weep: The Emotional Lives of Animals.* New York: Delacourte, 1995.

Midgley, Mary. *Evolution as a Religion.* New York: Methuen, 1985.

Spencer, Herbert. *Social Statistics.* London: Chapman, 1851.

Stebbins, George Ledyard. *Darwin to DNA, Molecules to Men.* San Francisco: Freeman, 1982.

Taylor, Gordon Rattray. *The Great Evolution Mystery.* New York: Harper and Row, 1983.

Thich Nhat Hanh. *Peace is Every Step.* New York: Bantam, 1991.

Trobriand Islanders. In *Millenium: Tribal Wisdom and the Modern World,* by David Maybury-Lewis. New York, Viking, 1992.

Trungpa, Chögyam. *Introduction to Buddhist and Western Psychology.* Edited by Nathan Katz. Boulder: Prajna Press, 1983.

Turnbull, Colin. *The Forest People.* New York: Touchstone, 1961.

————. *The Mountain People.* New York: Touchstone, 1972.

Wilson, Edward O. *On Human Nature.* Cambridge: Harvard University Press, 1978.

## LETTER 13: SECOND INTERLUDE

Chödrön, Pema. *Start Where You Are: A Guide to Compassionate Living.* Boston: Shambhala, 1994.

Fryba, Mirko. *The Art of Happiness: Teachings of Buddhist Psychology.* Boston: Shambhala, 1989.

Hayward, Jeremy. *Sacred World: A Guide to Shambhala Warriorship in Daily Life.* New York: Bantam, 1995.

Sogyal Rinpoche. *The Tibetan Book of Living and Dying.* New York: Harper-Collins, 1992.

Trungpa, Chögyam. *Training the Mind and Cultivating Loving-Kindness.* Boston: Shambhala, 1993.

## LETTER 14: BOUNDARIES IN SPACE: THE STUFF THE WORLD IS MADE OF

Calder, Nigel. *Einstein's Universe.* New York: Penguin 1980.

Davis, Paul. *Superforce.* New York: Simon and Schuster, 1984.

Greenstein, George. *The Symbiotic Universe.* New York: Morrow, 1988.

Kaku, Michio, and Jennifer Tranier. *Beyond Einstein: The Cosmic Quest for the Theory of the Universe.* New York: Bantam, 1987.

Pagels, Heinz. *The Cosmic Code.* New York: Simon and Schuster, 1982.

Puthoff, Hal. In "Inertia: Does Empty Space Put up Resistance," by Robert Matthews. *Science,* February 4, 1994.

Rock-climber quote in *Flow: The Psychology of Optimal Experience,* by Mihaly Csikszentmihalyi. New York: Harper and Row, 1990.

Wang Shihuai. In *Awakening to the Tao*, by Liu I-Ming. Translated by Thomas Cleary. Boston: Shambhala Publications, 1988.


## Letter 15: Awareness, Space, and Energy Come Together at Last

Bohm, David. *Wholeness and the Implicate Order*. London: Routledge and Kegan Paul, 1980.

———, and F. David Peat. *Science, Order, and Creativity: A Dramatic New Look at the Creative Roots of Science and Life*. New York: Bantam, 1987.

Bohr, Niels. *Atomic Physics and Human Knowledge*. New York: Science Editions, 1958.

———. *Essays 1958–1962*. New York: Interscience, 1963.

Bruner, Jerome. Conversation with Niels Bohr, in *Thematic Origins of Scientific Thought*, by Gerald Holton. Cambridge: Harvard University Press, 1973.

d'Espagnat, Bernard. *The Conceptual Foundations of Quantum Mechanics*. Reading, Ma.: W. A. Benjamin, 1976.

Davies, P. C. W., and J. R. Brown, eds. *The Ghost in the Atom*. Cambridge: Cambridge University Press, 1986.

Davies, Paul. *Other Worlds*. New York: Simon and Schuster, 1980.

de Witt, Bryce, and Neil Graham. *The Many Worlds Interpretation of Quantum Mechanics*. Princeton: Princeton University Press, 1982.

Dong, Paul, and Aristide H. Esser. *Chi Gong: The Ancient Chinese Way to Health*. New York: Paragon, 1990.

Folse, Henry. *The Philosophy of Niels Bohr*. Amsterdam: North Holland, 1985.

French, A. P., and J. P. Kennedy, eds. *Niels Bohr: A Centenary Volume*. Cambridge: Harvard University Press, 1985.

Heisenberg, Werner. *Physics and Philosophy*. New York: Harper and Row, 1958.

Herbert, Nick. *Elemental Mind*. New York: Plume, 1994.

———. *Quantum Reality: Beyond the New Physics*. New York: Anchor Press, 1985.

Josephson, Brian, and Ramachandran, eds. *Consciousness and the Physical World*. Elmsford, N.Y.: Pergamon Press, 1980.

Pauli, Wolfgang. Quote is from George Wald, below.

Penrose, Roger. *The Emperor's New Clothes*. Oxford: Oxford U.P., 1995.

Sheldrake, Rupert. *A New Science of Life*. Los Angeles: J.P. Tarcher, 1981.

Wald, George. "The Cosmology of Life and Mind," in *Synthesis of Science and Religion*. Edited by T.D. Singh and Ravi Gomatam. San Franscisco: The Bhaktivedanta Institute, 1987.

Wigner, Eugene P. *Symmetries and Reflections*. Cambridge: MIT Press, 1970.

Wilbur, Ken. *The Holographic Paradigm*. Boulder: Shambhala, 1982.

Young, Louise B. *The Unfinished Universe*. New York: Simon and Schuster, 1988.

## LETTER 16: MIND AND MATTER JOIN IN THE WORLD SOUL

Hannah, Barbara. *Jung, His Life and Work*. Boston: Shambhala, 1991

Harpur, Patrick. *Daimonic Reality: Understanding Otherworld Encounters*. London: Arkana, 1995.

Hillman, James. *Re-Visioning Psychology*. New York: Harper Perennial, 1992.

Hillman, James, and Michael Ventura. *We've Had a Hundred Years of Psychotherapy and the World's Getting Worse*. New York: HarperCollins, 1992.

Jung, Carl G. *Memories, Dreams, and Reflections*. New York: Pantheon, 1961.

———. *Man and his Symbols*. New York: Dell, 1964.

———. *Synchronicity*. Princeton: Princeton University Press, 1973.

Peat, F. David. *Synchronicity: The Bridge Between Matter and Mind*. New York: Bantam, 1987.

Sardello, Robert. *Facing the World with Soul*. New York: Harper Perennial, 1992.

von Franz, Marie-Louise. *Psyche and Matter*. Boston: Shambhala, 1992.

## LETTER 17: SUGGESTIONS OF MIND IN SPACE AT PRINCETON

Abell, George, and Barry Singer. *Science and the Paranormal*. New York: Scribners, 1981.

Broughton, Richard S. *Parapsychology: The Controversial Science*. New York: Ballantine, 1991.

Eysenck, Hans, and Carl Sargent. *Explaining the Unexplained*. London: Weidenfeld and Nicolson, 1982.

Jahn, Robert, and Brenda Dunne. *Margins of Reality: The Role of Consciousness in the Physical World*. New York: Harcourt Brace Jovanovich, 1987.

Targ, Russell, and Hal Puthoff. *Mind-Reach*. London: Granada, 1978.

Taylor, Gordon Rattray. *The Natural History of Mind.* New York: Penguin, 1981.


LETTER 18: HOW SHALL WE LOOK AT THINGS NOW?

Crichton, Michael. *Travels.* New York: Ballantine, 1988.

Koestler, Arthur. *The Ghost in the Machine.* New York: Random House, 1976.

Korzybski, Alfred. *Selections from Science and Sanity.* International Non-Aristotelian Library, 1972.

Varela, Francisco, Eleanor Rosch, and Evan Thompson. *The Embodied Mind.* Boston: M.I.T. Press, 1992.

Wilber, Ken. *Sex, Ecology, Spirituality: The Spirit of Evolution.* Boston: Shambhala Publications, 1995.


LETTER 19: WHAT DOES IT *FEEL* LIKE TO BE A TREE?

Chalmers, David. "The Puzzle of Consciousness." *Scientific American*, December, 1995.

———. "Facing Up to the Problem of Consciousness." *Journal of Consciousness Studies*, Vol. 2, No. 3, 1995. Several subsequent issues of this journal are devoted to the "Hard Problem."

Lowe, Victor. *Understanding Whitehead.* Baltimore: Johns Hopkins University Press, 1966.

Nagel, T. *The View from Nowhere.* New York: Oxford U.P., 1986.

Palter, Robert. *Whitehead's Philosophy of Science.* Chicago: University of Chicago Press, 1960.

Varela, Francisco. "Neurophenomenology." *Journal of Consciousness Studies*, Vol. 3, No. 4, 1996.

Whitehead, Alfred North. *Science and the Modern World.* New York: Free Press, 1967.


LETTER 20: HOLONS JUMPING UP AND DOWN

Briggs, John, and F. David Peat. *Turbulent Mirror: Chaos Theory and the Sciences of Wholeness.* New York: Harper and Row, 1989.

Jantsch, Erich. *The Self-Organizing Universe.* Oxford: Pergamon Press, 1980.

Prigogine, Ilya. *Order Out of Chaos.* Boulder: New Science Library, 1984.

Letter 21: Patterns of Life and Circuits of Thought

Bateson, Gregory. *Mind and Nature: A Necessary Unity.* New York: Bantam, 1980.

———. *Steps to an Ecology of Mind.* New York: Ballantine, 1975.

Davies, Paul. *God and the New Physics.* New York: Simon and Schuster, 1983.

Devereux, Paul. *Re-Visioning the Earth: A Guide to Opening Healing Channels between Mind and Nature.* New York: Fireside, 1996.

Feinberg, Gerald, and Robert Shapiro. *Life beyond Earth.* New York: Morrow, 1980.

Lovelock, James E. *Gaia: A New Look at Life on Earth.* Oxford: Oxford University Press, 1979.

Maturana, Humberto, and Francisco Varela. *The Tree of Knowledge: The Biological Roots of Human Understanding.* Boston: Shambhala, 1987.

Nollman, Jim. *Dolphin Dreamtime.* New York: Bantam, 1990.

Sperry, Roger. *Science and Moral Priority.* Oxford: Blackwell, 1983.

Thomas, Lewis. *The Lives of a Cell.* New York: Bantam, 1974.

Thompson, William Irwin, ed. *Gaia: A Way of Knowing.* Great Barrington, Ma.: Lindisfarne Press, 1981.


Letter 22: Third Interlude

Guenther, Herbert V., and Chögyam Trungpa. *The Dawn of Tantra.* Berkeley: Shambhala Publications, 1975.

Trungpa, Chögyam. *Journey without Goal: The Tantric Wisdom of the Buddha.* Boulder: Prajna Press, 1981.

Tulku Rinpoche, Ugyen. *Rainbow Painting.* Kathmandu: Rangjung Yeshe Publications, 1995.


Letter 23: Joining Heaven, Earth, and Human

Becker, Alton. "Text-Building, Epistemology, and Aesthetics in Javanese Shadow Theatre." In *The Imagination of Reality: Essays in Southeast Asian Coherence Systems.* Edited by A.L. Becker and A.A. Yengoyan. Norwood, N.J.: Ablex Publishing Coorporation, 1979.

Briggs, John, and F. David Peat. *Turbulent Mirror: Chaos Theory and the Sciences of Wholeness.*

Ehrenzweig, Anton. *The Hidden Order of Art.* Los Angeles: California U.P., 1971.

Gleick, James. *Chaos, the Making of a New Science.* New York: Viking, 1987.

Keeler, Ward. *Javanese Shadow Plays, Javanese Selves.* Princeton: Princeton U.P., 1987.

Madelbrot, Beniot. *The Fractal Geometry of Nature.* San Franscisco: W.H. Freeman, 1982.

Peitgen, H.O., and P.H. Richter. *The Beauty of Fractals.* Berlin: Springer-Verlag, 1986.

Rai, Jacota. Quote in *Utne Reader*, July-August, 1996.

Snake Dance. In *Another Look in the Mirror,* by Trudy Sable. Halifax: Saint Mary's University MA Thesis, 1996.

Stewart, Ian. *Does God Play Dice? The Mathematics of Chaos.* Oxford: Blackwell, 1989.

Chinese rainmaker story in *Psyche and Matter*, by Marie-Louise von Franz. Boston: Shambhala Publications, 1992.

LETTER 24: BEING A TUNING FORK ON A COSMIC PIANO

Coe, Stella. *Ikebana.* Woodstock, N.Y.: Overlook Press, 1984.

Cowan, Eliot. *Plant Spirit Medicine.* Newbury, Or.: Swan, Raven & Co., 1995.

Fingarette, Herbert. *Confucius—the Secular as Sacred.* New York: Harper and Row, 1972.

George, James. *Asking for the Earth: Waking Up to the Spiritual/Ecological Crisis.* Rockport, Ma.: Element, 1995.

Ghiselin, Brewster. *The Creative Process.* New York: Mentor, 1952.

Harman, Willis, and Howard Rheingold. *Higher Creativity.* Los Angeles: Tarcher, 1984.

Herrigel, Eugen. *Zen in the Art of Archery.* New York: Vintage, 1971.

LaChappelle, Dolores. *Sacred Land, Sacred Sex.* Silverton: Finn Hill Arts, 1988.

Porter, Elliot, and James Gleick. *Nature's Chaos.* New York: Viking, 1990.

Roncalli, Lucia. "Standing by Process: A Midwife's Notes on Story-Telling, Passage and Intuition." In *Intuition: The Inside Story.* Ibid.

Sen XV, Soshitsu. *Tea Life, Tea Mind.* New York: Weatherhill, 1981.

Seng-ts'an. "On Trust in the Heart." In *Buddhist Texts Through the Ages*, edited by Edward Conze. New York: Harper and Row, 1954.

Spretnak, Charlene. *States of Grace: The Recovery of Meaning in the Postmodern Age.* New York: HarperCollins, 1991.

Stevens, John. *The Sword of No Sword.* Boston: Shambhala Publications, 1984.


## Letter 25: Our Never-Ending Story

Becker, Carl B. *Paranormal Experience and the Survival of Death.* New York: SUNY, 1993.

Dossey, Larry. *Meaning and Medicine.* New York: Bantam, 1992.

Grey, Margot. *Return from Death.* New York: Arkana, 1983.

Morse, Melvin. *Closer to the Light: Learning from the Near-Death Experiences of Children.* New York: Ballantine, 1990.

Ring, Kenneth. *Heading Towards Omega: In Search of the Meaning of the Near-Death Experience.* New York: William Morrow, 1985.

Wilson, Ian. *The After Death Experience.* London: Corgi, 1989.

# Index

# About the Author

JEREMY HAYWARD was born in England in 1940. He studied mathematics and physics at Trinity College, Cambridge University, receiving a Ph.D. in 1965. After his Ph.D., Jeremy apprenticed for nine months at the Medical Research Council's Laboratory of Molecular Biology in Cambridge, under the direction of Nobel Prize–winner Francis Crick. This was followed by four years of research in molecular biology at Massachusetts Institute of Technology and Tufts Medial School. He published professional papers in both physics and molecular biology.

During the research period, he became interested in consciousness and its relation to language, and began an intensive period of study of psychoanalysis, formal logic, and linguistics. He also worked with a group practicing the Fourth Way of G. I. Gurdjieff, a teaching on consciousness and its possible evolution, for people in ordinary life. Leaving research in 1969, Jeremy taught high school science for two years and took part in the development of new curricula for physics and elementary science.

In 1970, Jeremy met the visionary Tibetan Buddhist teacher Chögyam Trungpa, Rinpoche, and began scientific investigation of the nature of experience through the meditative practices of mindfulness-awareness, creative visualization, and rDzogs-chen. For three years he staffed Karmê Chöling, a then-fledgling Buddhist meditation center in Vermont. In the summer of 1974, he helped to establish the Naropa Institute in Boulder, Colorado, where he was vice president from 1975 to 1985, and remained a trustee until 1996. After Trungpa began to bring forth the spiritual teachings of Shambhala warriorship, in 1976, Jeremy worked closely with him to develop Shambhala Training, a nonreligious spiritual path for ordinary life.

In 1983, writing his first book, *Perceiving Ordinary Magic: Science and Intuitive Wisdom*, Jeremy began the difficult and sometimes painful personal task of bringing together the two sides of his life: his love of true science, and his longing for genuine wisdom. He believes that the most vital need for our society, for the earth, and for the education of our

young people is to come to a common view that encompasses science and spiritual wisdom. In 1987 he was one of a small group of scientists invited to meet privately with the Dalai Lama for a week to discuss the sciences of life and mind. The record of these meetings was published as *Gentle Bridges*. He is on the editorial advisory board of the *Journal of Consciousness Studies*.

Jeremy now lives in Halifax, Nova Scotia, with his wife, Karen, and daughter, Vanessa. He is education director of Shambhala Training International. In 1997, Vanessa is in her final year of high school, after which she will take some time to work and travel. She is interested in eco-agriculture and plans to go to college when, and if, it makes sense for her life's journey.